Human Evolution
AN ILLUSTRATED INTRODUCTION

Human Evolution

AN ILLUSTRATED INTRODUCTION

ROGER LEWIN

American Association for the
Advancement of Science

W. H. FREEMAN AND COMPANY

NEW YORK

Published in the United States in 1984
by W.H. Freeman and Company,
41 Madison Avenue, New York,
NY 10010. First published in Great
Britain in 1984 by Blackwell Scientific
Publications Ltd
Reprinted 1985 (twice), 1986

Printed in Great Britain at the
University Press, Cambridge

Library of Congress
Cataloging in Publication Data

Lewin, Roger.
    Human evolution.
    Bibliography: P.
    Includes index.

    1. Human evolution.   I. Title.

GN281.L49   1984   573.2   84-4188
ISBN 0-7167-1635-6
ISBN 0-7167-1636-4 (pbk.)

Contents

Preface

Palaeoanthropology—the study of human origins—has undergone a transformation in recent years. Not only has the science benefited from a remarkable series of fossil discoveries, but it has also begun to encompass radically new approaches to answering the major questions about human evolution. Moreover, there has been a dual shift in many researchers' perceptions: one represents a salutary recognition of the depth of ignorance under which the science still labours and of the real limitations to what can ultimately be known; and second, there is a greater security about genuine progress that has been made, specifically progress in asking answerable questions.

Unquestionably, it is an exciting time for palaeoanthropology, not least because the related pursuits of evolutionary biology and geology are both experiencing a degree of creative foment that from time to time occurs in the history of a science and raises it to new, high frontiers.

In evolutionary biology, there is keen debate over the circumstances under which new species arise, specifically about the process of speciation, through which biological diversity expands. This debate is forcing palaeoanthropologists to scrutinize long-held opinions on the tempo and mode of the emergence of *Homo sapiens* and its ancestral species. And in geology the traditional theme of steady, gradual change through geological time is being challenged by remarkable discoveries about the nature of mass extinctions. Geological history, and therefore the history of life on earth, is beginning to be seen as much more episodic and unpredictable than previously appreciated. This inevitably influences the larger context in which human evolution is envisaged.

Probably the most significant discovery of the century regarding human origins—or Man's Place in Nature, as Darwin's friend Thomas Henry Huxley put it—has concerned our relationship with the apes. Until biochemical studies in the 1960s showed it to be fallacious, it was widely held that the great apes (gorilla, chimpanzee and orangutan) were closely related to each other, while humans were separated by a significant evolutionary distance. As the protein biochemistry showed, and as the more recent work on the DNA behind it confirmed, humans and the African apes (chimpanzee and gorilla) are closely related to each other and as a group are relatively distant from the Asian ape (orangutan). *Homo sapiens*, in many respects, is simply a rather odd African ape.

The coming together of traditional palaeoanthropology and modern molecular biology has been one of the most significant developments in the quest for human origins, although the partnership has at times been viewed askance by both parties. The union promises to offer powerful tools for helping to establish the all-important timescales of human evolution.

Human Evolution: an Illustrated Introduction brings together the many different approaches—the established and the new—that constitute the modern science of palaeoanthropology. It introduces the reader to the full range of questions that is being asked about this rather odd African ape—ourselves. What made an ancient ape stand on two legs instead of four? What effect did the advent of stone tool technologies have on our ancestors' energy budget? What forces of natural selection favoured the remarkable expansion of the brain? Is language simply a super-efficient means of communication, or a medium of deeper thought and of shared consciousness? What did the cave art and carving of the ice age signify? Was it population pressure that ignited the agricultural revolution, or was it something more subtle? These, and many more, questions are what we need to know about ourselves and our ancestors.

This book is not meant as an exhaustive treatment of physical anthropology, geology, evolutionary biology, molecular biology or archaeology; it is an introduction to these subjects as they impinge on the story of human evolution. *Human evolution* therefore provides an effective means of gaining a wider perspective of these subjects, as well as offering the student of human origins a broad, up-to-the-minute overview of the state of this very exciting science.

Roger Lewin
Washington, D.C.

Acknowledgements

No one can write a book with so broad a scope as this without being indebted to scores of practitioners who, over the years, have been patient enough to discuss at length their specific concerns. It is surely impossible to list all by name, but I should like to mention the following, who helped either as friends or just friendly advisors: Peter Andrews, Anna K. Behrensmeyer, Barbara Bender, C.K. (Bob) Brain, Margaret Conkey, Michael Day, Niles Eldredge, Kent Flannery, Stephen Jay Gould, Andrew Hill, Ralph Holloway, Nicholas Humphrey, Glynn Isaac, David Jablonski, Donald Johanson, Clifford Jolly, Lawrence Keeley, William Kimbel, Kamoya Kimeu, Richard Klein, Misia Landau, Mary Leakey, Richard Leakey, Alexander Marshack, Robert Martin, Ernst Mayr, Theya Molleson, David Pilbeam, David Raup, Vincent Sarich, Pat Shipman, Phillip Tobias, Nicholas Toth, Erik Trinkaus, Elizabeth Vrba, Alan Walker, Tim White, Allan Wilson, Edward O. Wilson, Bernard Wood, Milford Wolpoff, and Adrienne Zihlman. My wife, Gail, has helped, as always, more than she will ever know.

R.L.

1/Human Evolution in Perspective

The species *Homo sapiens sapiens* is a relative youngster on this planet, having arisen perhaps 100 000 to 40 000 years ago; and the family to which it belongs, the *Hominidae*, has roots that reach back only five to ten million years. By comparison, life on earth has a history almost four billion years long.

One of the most remarkable things about life on Earth is the speed with which it arose following the planet's formation, 4.6 billion years ago. Direct evidence from microfossils and indirect inference from molecular data show that the first primitive organisms arose within a few hundred million years of the planet becoming cool enough to support life. Then, for the next two billion years, the most complex forms of life were algal mats growing in profusion in shallow tidal waters. A microbiological melee of blue–green algae and bacteria, these algal mats pumped oxygen into the primitive atmosphere in great quantities. This process induced changes in the biochemistry of life's simple organisms and produced monumental 'rusting' (the formation of iron oxide) through vast geological formations.

Around 1.5 billion years ago the next major evolutionary innovation occurred: eukaryotic cells arose, cells that packaged their genetic material within discrete nuclei and performed photosynthesis

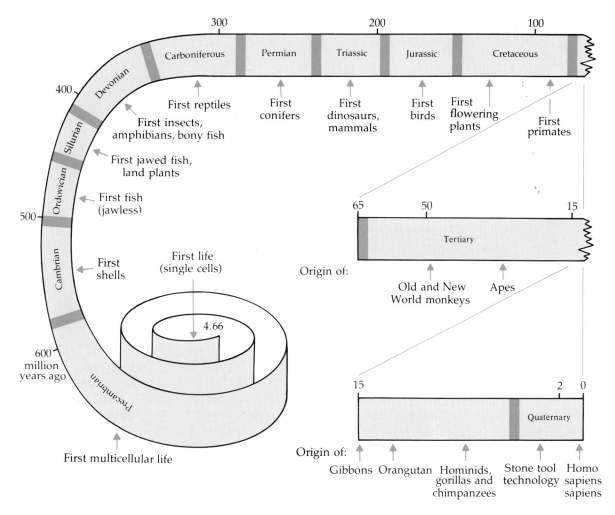

Evolutionary time scale. Although life arose very early in earth history, complex life forms are relative newcomers to the planet. Primates, the order to which *Homo sapiens* belongs, arose more than 65 million years ago. The first hominids, the name given to the human family, appeared perhaps as recently as five million years ago.

and respiration within chloroplasts and mitochondria, respectively. This new form of cell was capable of a new form of reproduction: sex, which allowed a new dimension of genetic variability within populations, a crucial raw material in the evolution of diversity.

diversification. First, a huge variety of mammal-like reptiles, large and small, herbivore and carnivore, dominated the world of terrestrial vertebrates. These immensely successful creatures were eclipsed 200 million years ago by the arrival of those equally prolific reptiles, the dinosaurs.

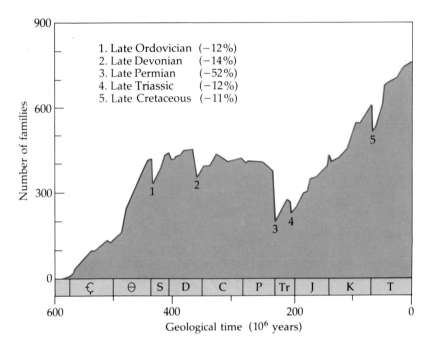

1. Late Ordovician (−12%)
2. Late Devonian (−14%)
3. Late Permian (−52%)
4. Late Triassic (−12%)
5. Late Cretaceous (−11%)

Episodic nature of life's history. Since the origin of multicellular organisms in the late Precambrian, life's history has documented a steady rise in diversity, as recorded here by the increase through time in the number of families of marine vertebrates and invertebrates. Interrupting this rise, however, have been a series of mass extinction events (numbered 1–5), which have reduced diversity by the figures shown in parentheses. Each extinction was followed by rapid radiations of new organisms. Courtesy of David Raup.

Before about 800 million years ago life, which was still confined to oceans, was exclusively represented by single-celled organisms. The origin of multicellular organisms at this point marked another major evolutionary step. And during the next 400 million years most of the 30 or so existing major body forms or phyla had appeared, plus many more that have since vanished. This Cambrian Explosion, as it has been characterized, probably reflects an unencumbered invasion of empty ecological niches.

Dry land remained untouched by life until about 400 million years ago, when plants, fungi, invertebrates and finally vertebrates became terrestrial. The first terrestrial vertebrates, amphibians, had to remain near water because their mechanism of reproduction, which involved eggs vulnerable to desiccation, demanded it. Around 300 million years ago, the origin of the amniote egg, secured in a tough shell, freed the amphibians' descendants, the reptiles, of their dependence on water. The conquest of dry land was complete.

The Age of Reptiles, which spanned a great swath of history from 300 million to 65 million years ago, witnessed two distinct waves of evolutionary

The end of the dinosaurs' reign, 65 million years ago, was sudden in geological terms, and is known as the Cretaceous extinction. However, their sudden demise was not especially cataclysmic from a biological perspective: their numbers declined over a five or ten million year period. Climatic changes were probably the principle cause of the dinosaurs' extinction, although there is persuasive evidence of asteroidal impact around 65 million years ago that might have delivered the *coup de grace*.

Before the Cretaceous extinction very few vertebrate species were smaller than a cat. After the extinction there was virtually no vertebrate larger than a cat. The living world had changed dramatically, and the Age of Reptiles had given way to the Age of Mammals.

Mammals have their origins some 200 million years ago, being descendants of the mammal-like reptiles, but it was not until the dinosaurs had vanished that they became abundant and began to occupy the large animal niches. Although they never matched the gargantuan size of some of the dinosaurs, some terrestrial mammals, from their diminutive origins, reached impressive proportions. The mammalian evolutionary odyssey, from tiny beginnings eventually to give rise to some very large species, is a common pattern repeated throughout the history of life.

Primates, the mammalian order to which humans belong, were probably already in existence when the Cretaceous extinction occurred: they were small, arboreal, nocturnal insectivores. Arboreality is a strong theme of primates, as are their grasping hands, excellent vision and high sociality. These small primates, the prosimians, gave rise to New and Old World monkeys some 50 million years ago. The apes arose 20 million years later in the Old World tropics, and became as diverse and abundant as monkeys are today.

Deteriorations in global climate from 20 million years ago onwards had a great impact on these large primates, creatures of the tropics. Many of these apes became extinct. Between ten and five million

Kingdom Animalia
Phylum Chordata
Class Mammalia
Order Primates
Family Hominidae
Genus Homo
Species Sapiens

Taxonomic classification of modern humans. Typically, modern humans are assigned to a further subdivision, the subspecies *Homo sapiens sapiens*, in order to separate us from a discrete, earlier and now extinct group, *Homo sapiens neanderthalensis*.

years ago one of them changed, giving rise to the direct ancestors modern African apes and humans.

The history of life is dramatically episodic, with great waves of extinction followed by bursts of adaptive radiation. Indeed, there is some evidence that major extinctions occur regularly – about every 26 million years – perhaps as a result of periodic impacts of comets or asteroids. Although the passage of time sees the origin of more and more complex forms, this is merely the outcome of the essentially random process of evolution operating on the material available at each stage. It is not in any sense a progressive programme of improvement.

If a glance at the history of life teaches anything, it teaches that every species will eventually become extinct. Today, the earth is populated by perhaps two million species, most of which are insects, a great proportion are plants, 8000 are amphibians and reptiles, 8600 are birds, and only 4000 are mammals. This two million represents just one per cent of all the species that have ever lived. The average 'life-span' of an invertebrate species is between five and ten million years, and for vertebrate species it is about one-half this or even less. Hominid species, according to the fossil record, survive for only one or two million years. *Homo sapiens sapiens,* the latest species on the hominid lineage, has existed so far for only 0.1 million years. The planet Earth can expect to continue for another ten billion years before the sun's hydrogen fuel becomes too depleted to support life on Earth.

2/Darwin and Natural Selection

In his most famous book, 'On the Origin of Species', published in 1859, Charles Darwin studiously avoided the issue of human evolution, but for one sentence in its conclusion: 'Light will be thrown on the origin of man and his history'. Darwin was principally concerned with presenting a persuasive case by which to establish the fact of evolution and with suggesting a mechanism, that of natural selection, by which organic change might occur through time. None of Darwin's many intellectual predecessors who broached the subject of evolution had managed to achieve this.

During his 40 000 mile, five year-long voyage around the world aboard HMS Beagle, the young Darwin observed the patterns of geology and of life, both extant and extinct, on many continents. What he saw impressed him greatly and was to become the core of the 'Origin'. However, as a keen pigeon fancier and lover of horses, Darwin was sensitive to the power for selective change of careful domestic breeding. And it was with an essay on 'variation under domestication' that he began his 'one long argument,' as he described the 'Origin'.

Extrapolating from the efficacy of domestication, Darwin argued as follows: 'Why, if man can by patience select variations most useful to himself, should nature fail in selecting variations useful, under changing conditions of life, to her living products. . . . I can see no limit to this power, in slowly and beautifully adapting each form to the most complex relations of life.' Unlike the conscious selection of the domestic breeder, however, natural

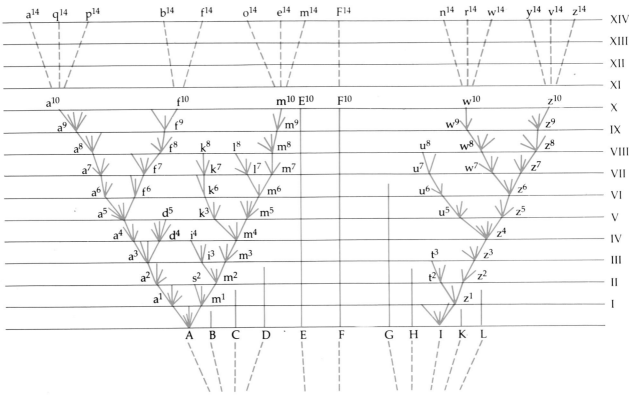

Darwin's only diagram in 'Origin of Species'. The capital letters A–L represent species in a genus, some of which changed through time (A and I), while the remainder produced unchanging descendants. The space between horizontal lines was meant to represent 1000 generations, so that after 10 000 generations species A had given rise to

three new species (a^{10}, f^{10}, and m^{10}), and after 14 000 generations to eight new species (a^{14}, q^{14}, p^{14}, b^{14}, f^{14}, o^{14}, e^{14} and m^{14}). Similarly, with species I. As species descendants of A and I proliferated they eclipsed the related species, most of which (except F) became extinct after 14 000 generations.

selection is not necessarily progressive and certainly not directed or purposeful. But neither is it totally random. In natural selection, genetic variation within a population, which is random, passes through the selective filter of environmental pressures, which is not random.

Darwin described the essence of natural selection as follows: 'More individuals are born than can possibly survive. A grain in the balance will determine which individual shall live and which shall die—which variety or species shall increase in number, and which shall decrease, or finally become extinct.' The nearly-perfect match between organisms and their way of life, viewed by natural theologians as the product of a Great Designer, is explained by natural selection as a passive adaptation through the differential survival of the fittest individuals.

If, for example, a dark coat colour confers a survival advantage in a population, perhaps through more effective camouflage, then those individuals with the darkest coats will have an improved chance of survival and, therefore, an improved chance of passing their genes to subsequent generations. Through time dark coats will predominate. And through long enough periods, given sufficient change through adaptation, new species will emerge, new forms of life will arise.

facit saltum', nature does not make jumps, he wrote in the 'Origin'. This, as will be seen later, has been a point of contention among evolutionary biologists for many years.

Darwin admitted that his theory had problems, but he adduced massive support for the fact of evolution. Drawing on his experience on the voyage of the Beagle, he noted that fossil successions throughout the world showed, in each case, a distinct continuity, a degree of relatedness over time not readily explicable by special creation. He noted, too, that living organisms within a class are plainly related to each other within discrete geographical areas. He noted that one should not be surprised that on continents that experience similar climatic conditions which are widely separated on the globe, the forms of life inhabiting them might be quite different: '. . . the course of modification in the two areas will inevitably be different.' Why should the patterns of life on oceanic islands be more similar to those on a neighbouring continent than to the patterns of life on ecologically similar islands elsewhere on the globe? he asked. 'It must be admitted that these facts receive no explanation on the theory of creation.'

Homologous structures—for example, the common pattern of bones in the human hand, the horse's leg, the bat's wing and the fin of a porpoise—also

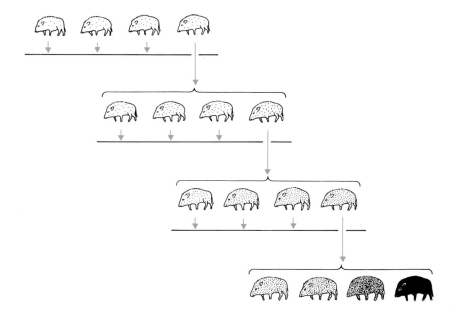

Natural selection. Selective advantages of large body size and dark coat colour confer differential survival and ultimate reproductive success on those individuals with those characters. As a result, the population mean for these characters will shift, generation by generation, towards larger size and darker coats.

The essence of natural selection as Darwin saw it was the steady accumulation of tiny incremental modifications—descent with modification as he called it. It is a slow and gradual process. 'Natura non

speak of descent with modification, Darwin argued. Descent with modification also explains why 'the embryos of mammals, birds, reptiles and fishes should be so closely alike.'

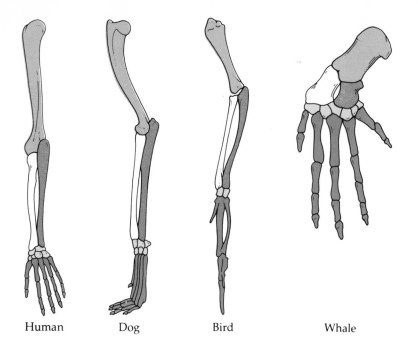

Human   Dog   Bird   Whale

The principle of homology. The biological derivation relationship (shown by colours) of the various bones in the forelimbs of four vertebrates is known as homology and was one of Darwin's arguments in favour of evolution. By contrast, the wing of a bird and the wing of a butterfly, although they do the same job, are not derived from the same structures: they are examples of analogy.

With surprising rapidity, Darwin's argument for the fact of evolution was embraced by the intellectual community. (The idea of natural selection was, however, the subject of debate many years afterwards.) With the intellectual climate so transformed, Darwin subsequently felt able to publicize his views on the more sensitive subject of human evolution. In 1871 he published 'The descent of man, and selection in relation to sex', which was an uncompromising discussion about man's place in the natural world.

'The main conclusion arrived at in this work, namely that man is descended from some lowly organized form, will, I regret to think, be highly distasteful to many', he wrote. 'We must, however, acknowledge, as it seems to me, that man with all his noble qualities, with sympathy which feels for the most debased, with benevolence which extends not only to other men but to the humblest living creature, with his godlike intellect which has penetrated into the movements and constitution of the solar system—with all these exalted powers—man still bears in his bodily frame the indelible stamp of his lowly origin.'

Like his great friend and champion, Thomas Henry Huxley, Darwin was greatly impressed by the anatomical similarities between humans and the African great apes, the chimpanzee and gorilla. From this, he speculated on the location of man's origins. 'In each great region of the world the living mammals are closely related to the extinct species of the same region. It is, therefore, probable that Africa was formerly inhabited by extinct apes closely allied to the gorilla and chimpanzee; and as these two species are now man's nearest allies, it is somewhat more than probable that our early progenitors lived on the African continent than elsewhere.' A century after this was written, the facts appear to support the great man's speculation.

1796. *Zoonomia* in which Erasmus Darwin discussed ideas on 'transmutation' that presaged much of Charles Darwin's theory.

1802. *Natural Theology* by William Paley who argued that the Natural World is the product of divine design, an argument that remained strongly held for half a century.

1809. *Philosophie Zoologique* by Jean Baptiste de Lamarck who proposed an internal force that drives organisms up an evolutionary ladder. Lamarckism remained strongly supported until the end of the century in some academic quarters.

1830–1833. *Principles of Geology* by Charles Lyell: the foundation of modern geology and the essential context to Darwin's evolutionary ideas.

1858. Joint paper by Darwin and Alfred Russel Wallace to the Linnean Society, the first public presentation of the theory of natural selection.

1859. *On the Origin of Species* by Charles Darwin.

1863. *Evidence as to Man's Place in Nature* by Thomas Henry Huxley, Darwin's friend of advocate, who outlined uncompromisingly what Darwin had only implied in the *Origin* about human history.

1866. *Versuche über Pflanzen-hybriden*, the description by Johann (Gregor) Mendel of his breeding experiments that revealed the principles of Genetics that were to become so important in providing a scientific basis for variation. Mendel's publication remained unnoticed for more than 30 years.

1871. *The Descent of Man* by Charles Darwin in which he stated bold and detailed evolutionary ideas applied to human origins.

1900. Mendel's work was rediscovered and described in publications by Hugo de Vries, Carl Correns and Erich Tschermak.

1901. *Die Mutationstheorie* by Hugo de Vries, a popular evolutionary theory that emphasized mutations over selection.

1937. "Genetics and the origin of species." Theodosius Dobzhansky presents the genetic input that would be so important in the emerging Modern Synthesis.

1940. *The Material Basis of Evolution* by Richard Goldschmidt, a major statement of the Mutation Theory.

1942. *Evolution: The Modern Synthesis* in which Julian Huxley outlined the marriage of Genetics and Ecology in establishing natural selection as the core of evolutionary change.

1942. "Systematics and the origin of species" by Ernst Mayr, who laid the foundations of many modern ideas on speciation.

1944. "Tempo and mode in evolution" by George Gaylord Simpson, one of the modern greats. Again, seeds of many current ideas are to be found here.

Major publications in the history of evolutionary biology.

3/Modern Evolutionary Ideas

The history of evolutionary ideas reveals how very complex the ultimate questions of biology are: 'How do species arise?' and, 'How is it that species are apparently so clearly suited to the demands of their daily lifes?' Ideas have come and gone, some to return later in refashioned form. Periods of consensus have settled a calm confidence over the field, subsequently to be replaced by a foment of new and reincarnated ideas in an atmosphere of excitement and sometimes heated debate. The 1980s is unquestionably a period of intense and healthy turmoil in evolutionary biology, with a melding of new ideas and new data generating a tremendous sense of creative tension amid urgent debate.

Although some nineteenth century scientists were immediately persuaded by Darwin's arguments that natural selection was the main agent of evolutionary change, substantial numbers clung to other ideas, including the Lamarkian notion of an internal driving force reflecting an organism's 'needs' directing that change. Then, when Mendelian genetics was rediscovered at the turn of the century, a new school emerged; these were the mutationists led principally by Hugo de Vries. Shifts in an organism's genetic constituents provided the major propulsion for evolutionary modifications, they argued, an idea later charicatured by the phrase 'hopeful monster'. The mutationists viewed evolution as proceeding by sudden bursts of change propelled by internal events, namely mutations. Effects of the external environment, via selection, were relegated, at best, to a minor role.

The early decades of this century saw evolutionary biology in some intellectual disarray. By the 1930s and 1940s, however, a consensus began to emerge, later to be known as the modern synthesis, a term coined by Julian Huxley, grandson of Thomas Henry Huxley. The modern synthesis, which was the product of a marriage of the rapidly maturing field of genetics with the more established ideas of selectionism, all in the context of population biology, brought a rare unanimity to this traditionally tumultuous branch of science. The principal architects of the modern synthesis were, with Huxley, Theodosius Dobzhansky, Ernst Mayr, Sewell Wright, Ledyard Stebbins, George Gaylord Simpson and Bernard Rensch.

At its most extreme, the modern synthesis viewed

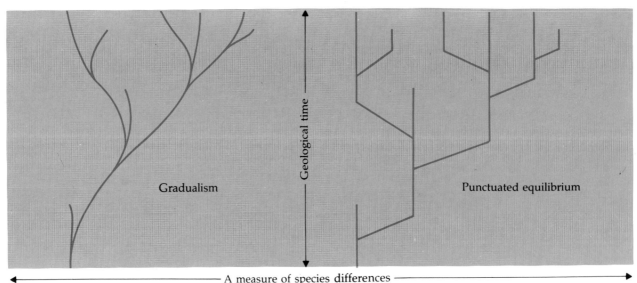

Gradualism

Punctuated equilibrium

Geological time

A measure of species differences

Two modes of evolution: gradualism and punctuated equilibrium. Gradualism views evolution as proceeding by the steady accumulation of small changes over long periods of time. Punctuated equilibrium, by contrast, sees morphological change as being concentrated in 'brief' bursts of change, usually associated with the origin of a new species. Evolutionary history is the outcome of a combination of these two modes of change; however, there is considerable debate as to which mode is the more important.

organisms as capable of infinite genetic (and, therefore, morphological) variation, with selection moulding a species from an infinite and continuous set of possibilities. Evolutionary change was characterized as a shift in gene frequencies, with traits being treated as discrete units. The overall effect of the modern synthesis was to rehabilitate the idea of evolutionary change as the product of the gradual accumulation of small incremental changes, the ultimate agent of change being natural selection. The implication was that the outcome of the all-powerful natural selection is near-perfect adaptation of species to their environments.

In 1972, two American palaeontologists, Stephen Jay Gould and Niles Eldredge, urged a reconsideration of saltational change—so-called punctuated equilibrium. The fossil record, for the most part, does not reveal a continuum of transitional forms between species. Each species in the record is relatively unchanging through time: it enters as the clear descendant of an earlier species; and when it disappears it is often replaced by a clear descendant of its own. Although such a chain of species may display undisputed relationship between one species and the next, typically there is a morphological gap in the fossil record between ancestor and descendant.

Darwin argued that the gaps resulted from an incomplete fossil record. Gould and Eldredge believe, incomplete though the record may be, it truly reflects the mode of evolutionary change: periods of morphological stasis are punctuated by bursts of change, or speciation events. These bursts might occur over 50 000 years, periods relatively long by biological standards but brief in a geological context.

Punctuated equilibrium is compatible with the mode of speciation championed since the 1950s by Mayr, that of allopatric speciation. New species typically arise, he says, in small isolated populations where 'genetic revolutions' are possible and new variants will not be diluted in a large pool of average genotypes.

Gould and others also object to the apparent implication of the modern synthesis that the range of variation available to selection is totally unconstrained and anything is possible. (The same observation had been made earlier by Conrad Waddington, but it was eventually submerged by the selectionism of the modern synthesis.) The counter argument holds that events of history and the rules of embryological development represent two important sources of constraint, and of opportunity, on evolutionary change. For example, the first terrestrial vertebrates had four legs, not because this was selected as the most efficient mode of locomotion but

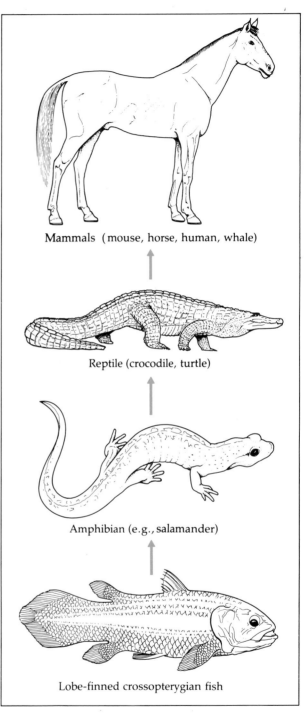

Mammals (mouse, horse, human, whale)

Reptile (crocodile, turtle)

Amphibian (e.g., salamander)

Lobe-finned crossopterygian fish

The principle of historical constraint. Evolution, in many ways, is a conservative process. The preservation of a four-limbed body over vast tracts of time and through very different environmental circumstances illustrates the power of historical constraint. For example, the horse has four legs not just because it is a very efficient way of moving about on dry land but because its fish ancestors also had four appendages.

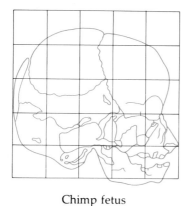

Chimp fetus

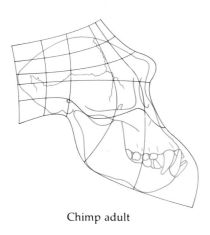

Chimp adult

Human fetus

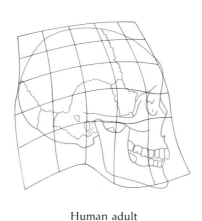

Human adult

Neotony in human evolution. Although the shape of the cranium in human and chimpanzee fetuses is very similar, a slowing down in development through human evolution has produced adult crania of very different forms, principally in the shape of the face and the size of the brain case. The changes in grid shapes indicate the orientation of growth.

because the first land animals descended from fish with four fins.

Although the rules of embryological development remain to be elucidated, it is clear that the embryo grows as an integrated whole. The fact that embryos of widely different organisms pass through very similar stages, as Darwin noted, seems to indicate that channels of development are indeed tightly constrained. The combined constraints of history and development must, therefore, limit the range of variation upon which selection can subsequently act.

The constraints of embryological development also present evolutionary opportunities. Small alterations in the timing of events in early development, for example, might produce a substantial change in the mature organism—not quite the 'hopeful monster' variety, but a more dramatic change than envisioned by a simple shift in gene frequencies. Indeed, there is a good deal of evidence that much of evolutionary change derives from shifts in the timing of developmental events, a pertinent example of

which is with humans. In many ways mature humans are reminiscent of juvenile apes: our small faces and globular cranium are examples of this. A crucial step in human evolution, the enlargement of the brain, can be seen as the result of a slowing down of embryological development in an ape-like ancestor. Instead of ceasing at birth, brain growth continues well into childhood, eventually producing a much larger and more complicated piece of mental machinery.

With natural selection remaining at the core of modern theory, constrained as it is by history and embryology, evolutionary change can be viewed as a combination of relatively rapid shifts and of gradual modifications. Major evolutionary innovations are likely to be the product of punctuational rather than gradual change. And there is decreased emphasis on viewing species as conglomerations of near-perfect adaptations. As French novelist Francois Jacob said: 'Evolution is a tinkerer, not a precision engineer.'

4 / Primate Heritage

Primates are quintessentially creatures of the tropics. *Homo sapiens,* having inhabited virtually every corner of the globe, is therefore something of an unusual member of this interesting vertebrate order. By contrast, many of the characteristics that might be taken to separate man from the rest of his fellow primates—such as extreme intelligence, upright walking, and intense sociality—are, in fact, merely extensions of typical primate features, not discontinuities from them.

The Cretaceous extinction that spelled the end of the reign of the dinosaurs also terminated many mammalian lines, particularly among the marsupials. Primates, in the infancy of their evolution at the time, were among the mammalian orders to survive the extinction. It is salutary to contemplate the course of history had the primate line been extinguished 65 million years ago! However, survive it did, and it experienced the kind of adaptive radiation typical of many mammalian groups through the Cenozoic period.

The earliest primates were small, nocturnal, arboreal animals, not unlike the modern tree shrew. Life in the trees is the natural habitat for the vast majority of primates and even those that have adopted a terrestrial life style, such as baboons and ring-tailed lemurs, are never far from the safety of branches aloft. It is not surprising, therefore, that adaptations to arboreality form the essence of what it is to be a primate.

The first primates were insect-eaters, a feature, incidentally, of many vertebrate orders in their evolutionary beginnings. For primates, the combination of predating on insects while suspended precariously on thin branches and twigs, led to a suite of important adaptations.

The predatory weapon was the hand, which developed a high level of manipulative facility. The hand eventually acquired an opposable thumb, which aided in grasping prey, and sensitive finger pads backed by nails rather than claws, which extended primates' exploratory dimensions in their

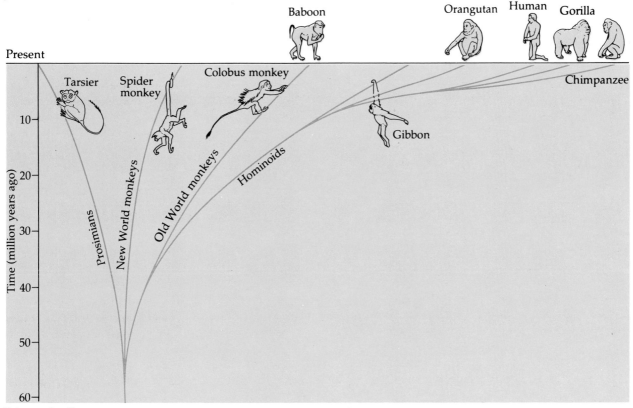

Primate family tree.

crepuscular world. The selective pressure for a keen ability for judging distance produced relocation of the eyesockets to the front of the head, which provided stereoscopic vision. The sense of smell, less useful in an arboreal environment than in the milieu of most mammals, diminished in acuity, particularly so in the higher, later evolved primates (the anthropoids: monkeys, apes and humans) as compared with the lower primates (the prosimians: lemurs,

Monkeys, by contrast, walk quadrupedally, whether along branches of trees or along the ground, supported on limbs of much more equal length. When at rest they sit in a relatively upright posture. Apes move beneath branches, rather than walking along them, suspended by long agile arms and shorter legs. Their hips became part of these creatures' vertical posture too. When they move across the ground, apes occasionally walk bipedally: it may be

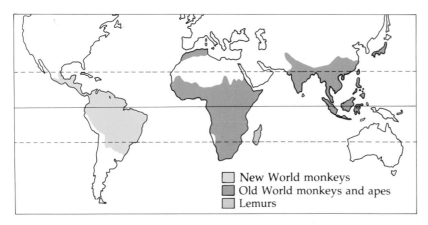

New World monkeys
Old World monkeys and apes
Lemurs

Geographical distribution of living primates.

lorises, tarsiers and sifakas, for example).

The adaptive radiation of the primates gave rise to larger species, the adoption of a diurnal life style, and the inclusion of plant foods (leaves and fruit) into the diet. The origin of monkeys and apes took these adaptations to ever higher levels. One consequence was the evolution of colour vision, which is now displayed by all primates, apart from the nocturnal prosimians and the South American night monkey. The possession of sensitive finger tips became of even greater importance, as the ripeness of fruits can, for the most part, be judged better by touch than by sight. There was a revolution in digestive anatomy and physiology too; an alimentary canal built to cope with the concentrated protein of insect bodies cannot digest the bulky cellulose of frugivorous or herbivorous diets. (In addition, leaf-eating often requires the ability to tolerate toxins.)

Although prosimians, monkeys and apes display three distinct modes of locomotion—known, respectively, as vertical clinging and leaping, quadrupedal walking, and brachiation—there is a common theme running through them that is pertinent to man's apparently special method of getting around. This theme is the persistent involvement of a relatively upright body.

The agile prosimians, powered by long muscular hind limbs, leap from tree to tree and spend much of their time clinging vertically to trunks or bows.

an awkward waddle, but it is bipedal nonetheless.

Uprightness, whether in clinging, sitting or walking, unquestionably is a feature of primate life, and it has impressed itself in the anatomy of the group. Changes in the skull and vertebral column have accommodated the demands of the upright posture. And the way the heart, lungs and viscera are suspended in the abdomen is different in primates from that in conventionally quadrupedal animals. Overall, then, the development of fully bipedal locomotion by early human ancestors may have been a less dramatic innovation than is often perceived.

One measure of the increase in complexity through evolutionary time has been an ever greater endowment of brain power relative to body size—called the encephalization quotient (EQ). Mammals, as a whole, have a greater EQ than reptiles, for example. And within the mammalian class, primates are undoubtedly the most intelligent order, having the highest EQ of all. Part of the enhancement of brain size and complexity in primates has to do with the nerve wiring and connections required for elaborated visual dependence and agility necessary for life aloft. However, one of the most complex aspects in the lives of all primates is social interaction, and this, unquestionably, has required greater mental agility than displayed in most mammals. Monkeys are more intelligent and more

Modes of primate locomotion. The monkey (top right) walks quadrupedally, while the gibbon (top left) is an adept brachiator (it swings from branch to branch like a pendulum). The orangutan (mid-left) is also adept in the trees, but as a four-handed climber. The gorilla (bottom left), like the chimpanzee, is a knuckle-walker (it supports the weight through the forelimbs on the knuckles of the hand rather than using a flat hand as the monkey does). The tarsier (foreground) moves by vertical clinging and leaping. The hominid (right) is a fully-committed biped. Note also, the grasping hands and forward-pointing eyes characteristic of primates. Courtesy of John Gurche/ Maitland Edey.

sociable than prosimians, and apes more so than monkeys. A concomitant of the quantity of learning demanded of intelligent animals is a lengthened period of infant dependency compared with that in most mammals.

The evolution of monkeys some 50 million years ago branched in two parallel directions, the New World and the Old World. The division, marked by some relatively minor anatomical differences, including the possession of a prehensile tail in the New World group, was a consequence of shifting continents which separated South America, as an island, from Africa. Apes evolved from the Old World monkeys, and by the Miocene, some 20 million years ago, had become extremely numerous and diverse. The Miocene, then, was the Age of the Ape. A resurgence of adaptive radiation among certain groups of monkeys, and a deteriorating climate, eventually eclipsed the ape's supremacy.

The lesser apes, the gibbon and the siamang, and the great apes, the orangutan, gorilla and chimpanzee, and humans are the modern and much-depleted representatives of a once widespread and extremely successful adaptation.

5 / Geology and Climate: background to Human Evolution

The oldest primate fossils to be discovered come from North America; specifically, they are teeth from a 65 million year old site at Purgatory Hill, Montana. As Montana can hardly be said nowadays to enjoy a tropical climate, the presence of primate fossils there seems at odds with the earlier assertion that primates are creatures of the tropics. Moreover, the fact that *Purgatorius*, as the Montana primate is known, lived in North America may also seem curious since when *Homo sapiens* entered the continent some 40 000 years ago there were no native primates in the land.

The history of life on earth is influenced in no small measure by the history of the continents. The light continental rock rests on the Earth's thin crust, which is composed of a dynamic mosaic of interlocking plates rather than being a continuous skin. The plates are in a perpetual state of formation or destruction at their various margins, and thus they move relative to each other, powered by convection currents in the liquid mantle. As a consequence, the continents, which ride as passive passengers on the plates, move too, occasionally colliding with each other, occasionally being torn apart.

The periodic coming together and separation of land masses affects not only global geography, which determines communication between the world's biotic communities, but also modulates global climate through altering atmospheric and oceanic circulation. Shifts in global climate inevitably leave their mark on the pattern of life's history.

Some 225 million years ago the world's land masses coalesced into a single equatorial super continent, Pangea. The formation of Pangea, incidentally, coincided with, and was almost certainly responsible for, the largest mass extinction in the history of our planet, the Permian extinction. The moving plates that had brought the continents into temporary union eventually began to tear them apart again. By 65 million years ago North America, Europe and Asia were loosely associated as the northern supercontinent, Laurasia. Meanwhile South America, Australia, India and Antarctica formed a fragmenting supercontinent, Gondwanaland, to the south. Since that time, the continents as we know them have drifted away from the equator, and the Americas in their westward journeys united only three million years ago.

So, *Purgatorius* did enjoy a warm climate in Montana 65 million years ago, but this did not last. Primates throughout the continuous land mass formed by North America and Eurasia more and more became restricted to southern latitudes as the continents drifted northward. The dependence of primates on tropical climes derives not so much from a need for warmth as for the year-round availability of succulent leaves, fruit and insects.

The precise location of the origin of primates remains to be established, but North America is an obvious possibility. Nevertheless, primates also lived in the southern continents early in the Cenozoic, specifically in Africa and South America. The second great phase of primate evolution, the origin of the monkeys around 50 million years ago, is also recorded only sketchily in the fossil record.

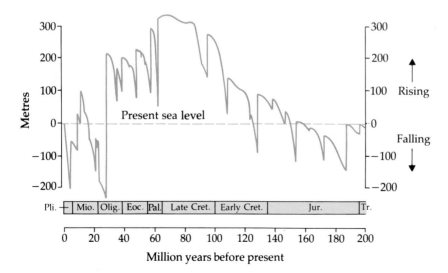

Sea level fluctuation. Sea level has fluctuated substantially over the past 200 million years, with one large cycle of rise and fall punctuated by short, rapid cycles. At least some of the precipitous drops are associated with the onset of glaciation, particularly in recent times (30 million years onwards). Adapted from Vail P.R., Mitchum R.M., Todd R.G., Widmier J.M., Thomson S. III, Sangree J.B., Bubb J.N. & Hatelid W.G. (1977) *Seismic-Stratiography Applications to Hydrocarbon Exploration.* (Ed. Payton C.E.) pp. 49–212. AAPG Memoir 26.

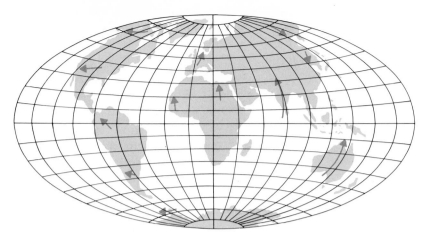

Present

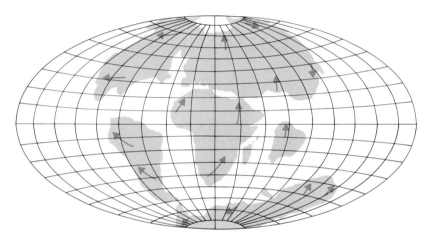

65 million years ago

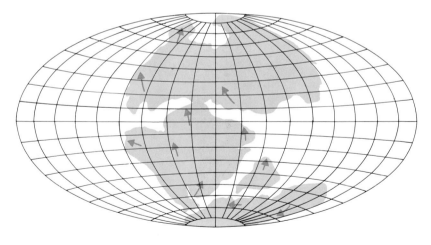

180 million years ago

Palaeogeographies as a result of continental drift. The bottom diagram shows the globe shortly after the supercontinent, Pangea, began to fragment. The migration of the continents through the ages has a profound influence on global climates. Arrows indicate direction and magnitude (length of arrow) of continental movement.

Although the New World and Old World monkeys are likely to have derived from a common stock, for the greater part of their evolutionary history they have remained isolated from each other as South America drifted westward away from Africa.

Apes appear to have originated around 30 million

15

years ago in Africa, perhaps in the north, while the continent was separated from Eurasia. The favourite candidate for the first ape is known as *Aegyptopithecus*, a cat-sized creature whose fossils have been found in the Fayum Depression in Egypt. The area is desert today, but 30 million years ago it was carpeted with thick forest. When Africa rejoined with Eurasia some 18 million years ago, there was an inevitable exchange of fauna between the two land masses, and apes quickly flourished throughout Southern Eurasia. The origin of the gibbon and orangutan lineages occurred close to this time, and may well have been related to this palaeogeographical event.

From about 20 million years ago, that is, early in the Miocene, the global temperature, which had been cooling for some time, experienced a sharp

east into rain shadow, which caused greater shrinkage of forests there. The upshot of all these changes was a replacement of what was once more or less continuous forest by a complex mosaic environment composed of some forest, woodland and open country. A terrain that formerly was ideal for apes steadily became less hospitable for them. Their numbers began to decline.

Towards the end of the Miocene, between seven and five million years ago, an abrupt drop in global temperature coincided with the formation of the West Antarctic ice sheet. Sea levels fell, the Mediterranean all but dried up, the tropical belt shrank even further and there were widespread extinctions of primates throughout Eurasia. Grasslands expanded in Africa, and this was accompanied by the flourish-

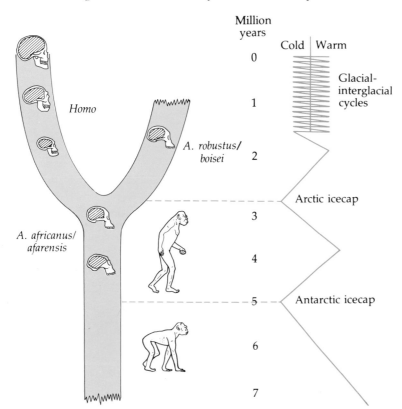

C.K. Brain's correlation between global low-temperature events and critical phases in hominid evolution. Courtesy of C.K. Brain.

deterioration and became more seasonal. As a result, the wide belt of tropical forest began to shrink throughout the Old World. Meanwhile, the same kind of tectonic activity that was moving the continents around the globe began to strain the crust beneath Eastern Africa. This produced a tremendous uplift in Kenya and Ethiopia, which gave rise to substantial highlands in those countries. Eventually, the strain on the continental rock proved too much, and the Great Rift Valley began to form. Meanwhile, the presence of the highlands threw the land to the

ing of swift plains animals equipped with teeth capable of grazing on tough grass. This also was the time, according to some estimates, when the final common ancestor gave rise to hominids on the one hand and the ancestors of the African great apes on the other. The first hominids were ape-like creatures that walked bipedally and lived in woodland rather than forest, but not in open savannah.

A second sharp deterioration in global climate occurred two and a half million years ago, when the northern ice-cap formed for the first time. This was

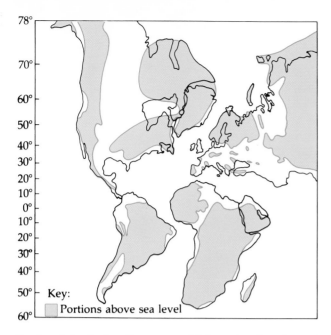

Sea levels in the late Cretaceous. High sea levels at the end of the Cretaceous 'drowned' large areas of the continents. Note the fragmentation of North America in to two land masses.

again a period of rapid evolution among animals adapted to living on the open plains. And, according to some opinions at least, it was also a time of important differentiation among the hominids, including the first appearance of the genus *Homo*.

During the past two million years the world has experienced more or less regular cycles of glaciation, each lasting around 100 000 years, each of which must have had a profound impact on the daily lives of our ancestors and some of which probably had an important influence on the course of their evolution. Currently, we are basking in the (probably) temporary respite of an interglacial.

Geology, climate and human evolution are closely linked, as C.K. Brain has recently emphasized. 'It is probably reasonable to conclude that, had it not been for temperature-based environmental changes in the habitats of early hominids, we would still be secure in some warm hospitable forest, as in the Miocene of old, and we would still be in the trees.'

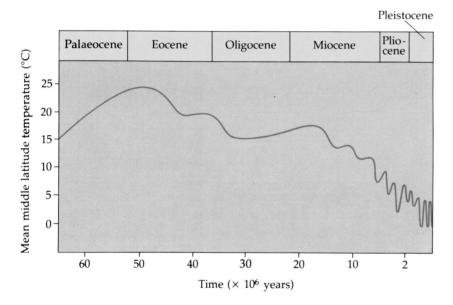

Global temperature changes.

6 / Molecules tell Evolutionary Stories

When Carolus Linnaeus drew up his classification of living organisms in the mid-eighteenth century, he recognized the many physical characteristics shared by *Homo sapiens* and the great apes. The same physical affinities led Huxley, Darwin and other evolutionists of the late nineteenth century to conclude that humans and apes descended from a common stock. Ironically, many scientists thereafter became uncomfortable with the extreme intimacy implied by these conclusions and made strenuous efforts to put as much evolutionary distance and time as possible between humans and apes. A decade ago, for example, most palaeoanthropologists defended the notion that the split between *Homo sapiens* and the African apes occurred as much as 20 million years ago.

During the past ten years there has been a dramatic

reduction in that all-important figure for the ape–human divergence: it is now thought to be much closer to five million years. Although fossil discoveries have played a crucial role in this shift of opinion, contributions from biochemistry and molecular biology have been a central factor, though often ignored until quite recently. Molecules have been important in providing 'clocks' by which to put a date on the divergence.

The idea of using molecules as phylogenetic clocks rests on one simple assumption: once two species separate in evolution the genetic material (DNA) in the two lines accumulates changes or mutations. The longer the separation time, the greater will be the sum of accumulated mutations. If the rate of accumulation remains steady through the ages—and there is a good deal of discussion on this point—a measure of the biochemical differences between the two species can be converted into a measure of the time since they derived from a common ancestor.

Morris Goodman, of Wayne State University, kindled modern interest in molecular clocks when in 1962 he published data on the immunological properties of the protein albumin, which showed chimpanzees, gorillas and humans to be closely

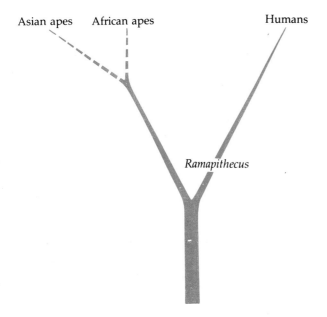

Premolecular evidence: Asian and African apes thought to be closely related; hominids diverged from apes prior to 15 million years ago, because of position of *Ramapithecus* as putative early hominid.

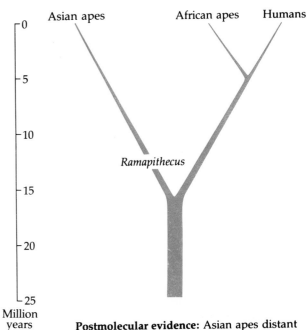

Postmolecular evidence: Asian apes distant from African apes, which diverged from hominids very recently (perhaps five million years ago). *Ramapithecus* cannot, therefore, be a hominid because it lived prior to the African ape/hominid divergence.

A scheme showing envisaged branching patterns, based on (1) premolecular evidence and (2) postmolecular evidence.

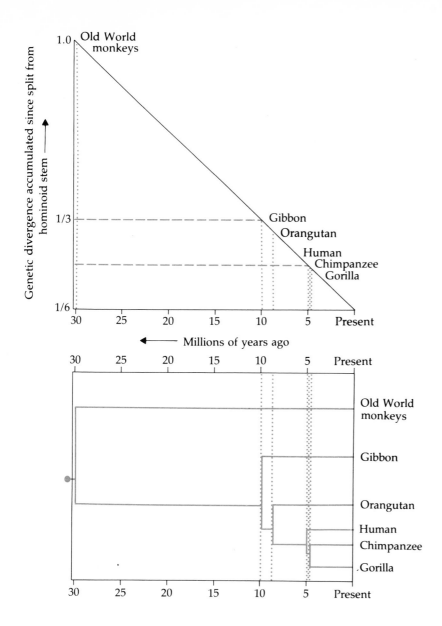

Genetic divergence accumulated since split from hominoid stem →

1.0 Old World monkeys

1/3 — Gibbon
Orangutan
Human
Chimpanzee
Gorilla

1/6

30   25   20   15   10   5   Present

← Millions of years ago

30   25   20   15   10   5   Present

Old World monkeys

Gibbon

Orangutan

Human

Chimpanzee

Gorilla

30   25   20   15   10   5   Present

A chart showing Wilson and Sarich's branching pattern derived from the genetic distance data. Using Old World monkeys as standards, Sarich and Wilson measured the genetic distance in African apes, Asian apes and humans and derived divergence times. From this, they inferred a branching pattern that grouped the Asian apes as having split away some ten to eight million years ago. The African apes and humans formed a second discrete group that diverged only five million years ago. Although early work could not separate the branching order between chimpanzees, gorillas and humans, more recent data indicate that the African apes briefly shared a common ancestor after diverging from the ancestor shared with hominids. The demonstration that humans group genetically very closely with the African apes and distantly from the Asian apes is one of the greatest discoveries of the century relating to human origins.

related to each other while the gibbon and orangutan were more distant cousins from this trio. According to the comparison of albumin from these three species, humans are as close genetically to chimpanzees and gorillas as these two apes are to each other. This was quite a shock, as chimpanzees and gorillas are morphologically rather similar to each other and to the orangutan, whereas humans seem unquestionably distinct from them.

Two biochemists at the University of California at Berkeley, Vincent Sarich and Allan Wilson, then published a landmark paper in 1967 that put a date on the chimpanzee–gorilla–human divergence. Again, using immunological properties to measure the differences in structure between the same protein

from the three species, Sarich and Wilson concluded that the African great apes and humans last shared a common ancestor five million years ago. The resolution allowed by the technique was insufficient to determine whether the divergence was a three-way split or, for example, the two apes shared a common ancestor briefly following the separation of the hominid line.

Sarich and Wilson's conclusion was not embraced with enthusiasm by the palaeoanthropological community. At the time, the community's favourite candidate for the first hominid in the fossil record was *Ramapithecus*, specimens of which from Asia, Europe and Africa showed that it lived at least 14 million years ago. A divergence between apes and

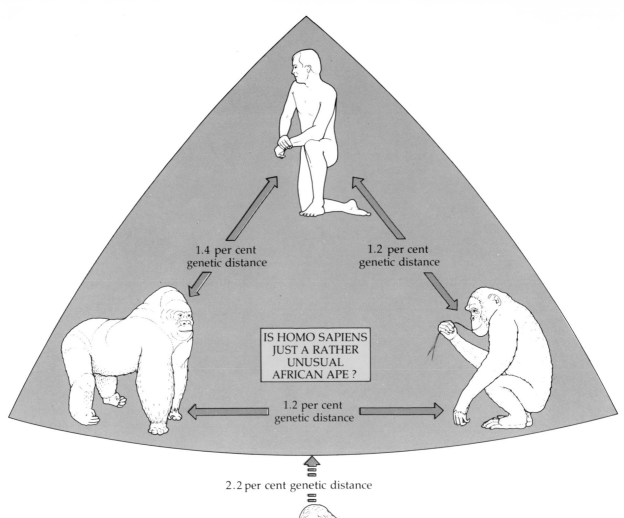

1.4 per cent
genetic distance

1.2 per cent
genetic distance

IS HOMO SAPIENS
JUST A RATHER
UNUSUAL
AFRICAN APE ?

1.2 per cent
genetic distance

2.2 per cent genetic distance

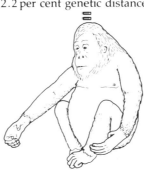

A chart to show genetic distance
between humans and the apes.

Species compared	% DNA difference
Human–gorilla	1.4
Human–chimpanzee	1.2
Gorilla–chimpanzee	1.2
Human–orangutan	2.4
Gorilla–orangutan	2.4
Chimpanzee–orangutan	1.8

Relationship between *Homo sapiens* and the African apes as
defined by molecular biology.

humans just five million years ago, as Wilson and Sarich contended, was therefore considered far too recent.

In the 15 years following the publication of Wilson and Sarich's provocative paper, immunological data on at least six independent proteins has been accumulated, and each appears to tell the same story. Many new potential clock techniques have been developed over the years, some in the Berkeley laboratory, others elsewhere. Some, such as the analysis of the nucleotide sequence of segments of DNA, have the advantage of giving a much greater resolution of genetic differences than is possible from the gross protein structure techniques of the original immunological method. Others, such as the physical matching of one strand of DNA with a counterpart from another species (DNA hybridization), attempt to average out complicating differences that arise from the exceedingly complex nature of genes and other genetic elements in the DNA. DNA hybridization, principally through the development of Charles Sibley and Jon Ahlquist at Yale University, is emerging as the most powerful of all the potential molecular clocks.

The simple assumption of a regularly ticking molecular clock turns out to be an over-simplification. A species' DNA is a patchwork of many different types of sequences, each of which might be susceptible to change in different ways. Ironically, however, inspite of this tremendous complication there *are* clock-like features in genetic change through time, even though they cannot yet be explained or modelled. The clock may be sloppy, but it tells the time nevertheless. The upshot of all this is that there remains the expectation that the molecules will be able to pinpoint the branch times in human history, but there is still a degree of disagreement over the data that are already available and their interpretation.

Wilson and Sarich, for example, argue that the conclusions from all the various molecular techniques consistently indicate five million years (plus or minus one million years) as the ape–human branch point. Others point to the DNA hybridization data in particular and claim a somewhat earlier divergence, between ten and seven million years. There is, however, consensus on the overall sequence of events: the gibbon diverged first, followed quickly by the orangutan; much later, the gorilla–chimpanzee–human split occurred. With the greater time resolution allowed by the modern molecular clock techniques it now begins to appear more likely than not that the gorilla split off first, leaving the chimpanzee and human lines briefly to share a common ancestor, before finally diverging.

Meanwhile, the palaeoanthropological community has been reassessing its position. Fossil finds of recent years have dislodged *Ramapithecus* from its status as the putative first hominid and much more recent divergence data—close to Sarich and Wilson's original five million year figure—are now being considered.

The small degree of genetic distance that separates *Homo sapiens* from the African apes—just one per cent in the genes that code for proteins—is the same as that often recognized in sibling species, that is, species that are barely separate in evolutionary terms. And yet humans and apes are assigned to different families, a much higher taxonomic division than species or even genera. Pressure is rising to address this issue and to recognize that, in many ways, *Homo sapiens* is really just a rather unusual African ape.

7/Major Steps in Human Evolution

One task of palaeontologists is to try to understand the social, economic and anatomical adaptations of each of the hominid species. Another is to seek some insight into the shift in adaptations that from time to time was associated with new species.

No other primate habitually walks upright, and in many ways the adoption of bipedalism can be seen as *the* crucial step in human history, a step that made everything else possible. A permanent upright stance freed the hands from locomotor function and allowed the development of fine manipulative skills. Stone tool technology, a consequent broadening of diet (to include meat), a more complex social structure, a blossoming of intelligence, all can be seen as a product, however distant, of this unusual posture. Evidence of bipedalism in the anatomy of hind limbs and pelvis and in 'fossilized' footprints is among the oldest hominid material discovered so far. The skeletal material comes from the Hadar region of

In looking back through human history in search of the principal evolutionary changes that eventually built *Homo sapiens,* it is easy, but wrong, to think of ancestral hominids as rather quaint but diminutive versions of our modern selves. Each species in the hominid lineage was a creature well-adapted to exploiting a certain type of existence, each a success story in its own right. Some of these species eventually became extinct without issue while others gave rise to descendants, one of which is *Homo sapiens.*

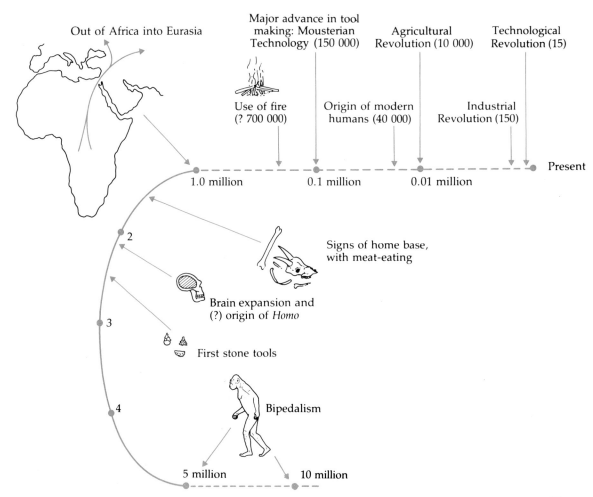

The major steps in human evolution. The origin of bipedalism and the subsequent expansion of the brain must rate as the two major milestones in human prehistory. The log of important changes in human evolution fills at an ever increasing rate through time.

Ethiopia and the footprints from Laetoli in Tanzania. The bones are between 3.0 and 3.6 million years old while the prints were made in volcanic ash 3.75 million years ago.

Clearly, bipedalism was an ancient hominid adaptation, perhaps the first hominid adaptation. Significantly, the earliest signs of tool technology and of an expansion of brain size appear much later in the record, close to two million years ago. One cannot, therefore, link these 'human' attributes with the initial evolution of upright walking.

The common ancestor of humans and the African great apes very likely divided its time between clambering in trees (brachiation) and moving across the ground, a typical ape-like existence. With the cooling climates and changing rainfall patterns in Africa five to ten million years ago, the environment enjoyed by this ancestral ape would have altered, principally in the diminution of forest cover and the spread of woodland and grassland. Apes would have been forced to retreat with the forests or adapt, through natural selection, to the new conditions. Chimpanzees and gorillas probably 'chose' the first option, with chimps subsequently moving back to more open woodland. The characteristic knuckle-walking style of these apes might have evolved in parallel in the two lines, or this form of locomotion might have been present in the common ancestor of apes and humans. Most anthropologists favour the former notion.

Meanwhile, the hominid line apparently took a different turn. It developed a mode of foraging for food (succulent shoots and tough fruits) more widely dispersed than is typical for apes: that of bipedalism. Given a tree-climbing ape as our ancestor, it can be shown that bipedalism is a satisfactory energetic solution to the need routinely to cover relatively large territories while foraging for food. Bipedalism is more efficient than, for example, the chimpanzees' knuckle-walking under these circumstances. In other words, the greatest evolutionary step of humans might well have been an anatomical change that allowed an ape to stay and forage in a terrain not really fit for an ape.

There were changes in the teeth and jaw structure in these early stages too, but nothing as radical as the advent of bipedalism. And nothing else very radical occurred in human prehistory for a very long time thereafter; not until some time between two and three million years, when, as mentioned earlier, the climate experienced another sharp deterioration. By this time there were several species of hominid living

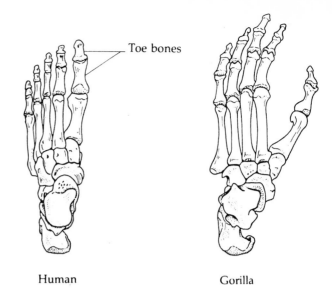

Top views of the left foot of a human and a gorilla. The opposable great toe in the gorilla, which is used in climbing, has become part of a platform structure with the lateral toes in humans.

All the fossil hominid feet so-far discovered fit closely with the human pattern, although those of the earliest, the 3.5 million year old *Australopithecus afarensis*, do have curved toes like those of an ape, which may indicate a residual climbing adaptation. The *afarensis* foot does not, however, have an opposable great toe; and the joints between the toes and the foot in *afarensis* allow for the same type of movement as in humans but different from apes.

in Africa. One was a robust creature, aptly named *Australopithecus robustus*. Another was a smaller model of robustus, *Australopithecus africanus*. In many ways, these two species were simply larger versions of the ancestral hominids from Ethiopia. A third, named *Homo habilis*, was, however, a distinct departure. This creature marked the beginning of significant brain expansion in the hominid lineage. Coincident with the appearance of *Homo habilis* in the fossil record is the beginning of the archaeological record: crude stone tools, pebble choppers and small sharp flakes are found, sometimes in association with animal bones. Evidently, the expansion of the brain and the invention of stone tool technology were accompanied by a change in diet, specifically the occasional inclusion of meat. There is, however, no evidence at this stage of regular full-scale hunting.

Further brain expansion marked the arrival of a new hominid species, presumably the successor to *Homo habilis*, some 1.5 million years ago: *Homo erectus*. Archaeologically, the record becomes more complex and there are persuasive examples of home

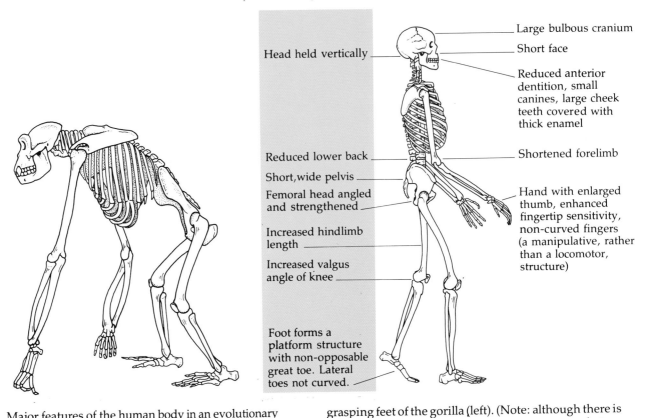

Head held vertically

Large bulbous cranium

Short face

Reduced anterior dentition, small canines, large cheek teeth covered with thick enamel

Reduced lower back

Shortened forelimb

Short, wide pelvis

Femoral head angled and strengthened

Hand with enlarged thumb, enhanced fingertip sensitivity, non-curved fingers (a manipulative, rather than a locomotor, structure)

Increased hindlimb length

Increased valgus angle of knee

Foot forms a platform structure with non-opposable great toe. Lateral toes not curved.

Major features of the human body in an evolutionary context. Much of man's lower body anatomy has been shaped by the bipedal style of locomotion; compare this with the long pelvis, relatively short hindlimbs and grasping feet of the gorilla (left). (Note: although there is some debate on the matter, the common ancestor between humans and the African apes is generally considered not to have been a knuckle-walker.)

bases, places to which food was brought, shared and consumed; this is a very un-ape-like behaviour. Meat apparently became a much more important part of these creatures' diet, although it is probably a mistake to think of *Homo erectus* as principally a hunter. Archaeological evidence of a preoccupation with hunting comes much later in the record. *Homo erectus* was the first hominid to venture out of Africa, perhaps because of its sharper wit, but, more likely, because the more eclectic diet allowed for the exploitation of environments much more varied than that typical of most primates.

The origin of *Homo sapiens* is difficult to pin-point in the record, but it must have been somewhere around 250 000 years ago, with truly modern humans perhaps 100 000 years ago or less. Again, the principal feature in the evolutionary change was further brain expansion, although dentally this species is much less robust than its forebears, as is its overall frame. These later stages of human prehistory are also characterized by much finer, more complex and quickly changing tool technologies, and, from about 30 000 years onwards, art, both painted and sculpted, appears in the record.

The two principal moving forces in human history, then, are upright walking and brain expansion. The impact of the latter has, of course, reached far beyond the well-being of the hominid line.

Expansion of mental capacity underlies all that is manifested in a highly intelligent, highly inventive, highly social creature. The development of a spoken language must have been the key event in this network of self-feeding factors. Language is of course invisible in the archaeological record. As invisible in the record is a human quality that transcends pure intelligence and in some ways is the greatest evolutionary innovation of them all: consciousness, the true questioning consciousness of self and of others that makes humans ask the question, Why? Glowing as an ember in the expanding cerebral cortex of our ancestors, this most intangible of all human characteristics has burst into a dazzling flame that illuminates what it is to be human: a cultural, ethicizing animal.

24

8/Past Ideas on Human Origins

Given the exquisite sensitivity of the topic, it is perhaps not surprising that the study of human origins, palaeoanthropology, is more emotional, more intense, and more given to flights of fancy than are most sciences. Although no science is the cool objective Baconian pursuit often idealistically portrayed, palaeoanthropology has been characterized by a distinctly greater-than-average level of eccentricity and private and public squabbles. And, just as any science reflects to some extent the current interests of the society in which it is being practised, so too has palaeoanthropology manifested passing fashions that can be traced to societal fads.

One very practical matter that has allowed and encouraged to an important extent the above-mentioned characteristics has been, until very recently, the virtual absence of data on which to construct theories. Until 1924 no hominid fossil older than about one million years had been discovered. Indeed, at the time, much of the palaeoanthropological community, especially in the United Kingdom, was happily blinded to reality by the seductive features of Piltdown Man, a brilliantly executed and eagerly accepted fraud. (Piltdown was not revealed as a fraud until 1953, but by then it was viewed by many with a very sceptical eye.) South

Africa was the principal source of early hominid fossils from 1924 until the 1960s, when the focus began to switch to East Africa. Sites at Olduvai Gorge and Laetoli in Tanzania, East Turkana in Kenya and most recently the Hadar region of Ethiopia have now yielded a substantial array of fossils that allows a degree of informed speculation about human history back to about four million years. Then, there are fossil accumulations in Pakistan, Turkey, China and elsewhere in Eurasia and again in Africa that sketch in events from eight million years back to 30 million years or more. There remains a glaring gap, between four and eight million years ago, that is essentially fossil-free. A pity, because this appears to be the time in which the most interesting evolutionary events took place.

Much of the older scientific literature on human evolution was concerned with the 'transformation' of ape into man, the ascent from the depths of bestiality to the heights of human civilization. Misia Landau, an anthropologist at Boston University, concludes from a study of the literature that such accounts of human origins, instead of being scientific constructions based on factual foundations, are more strongly influenced by the literary tradition of the 'hero myth'. They are literary narratives, not scientific theories. A hero (our ape-like ancestor) is introduced, is faced with certain tests (adverse climates and environments), and is eventually triumphant (reaches the status of *Homo sapiens* cradled by civilization). It is an account of triumph over adversity. The tradition has not lost its grip completely even today, suggests Landau.

In spite of their heretical message, Landau's argu-

In the early decades of the twentieth century two opposing views of human origins were current:

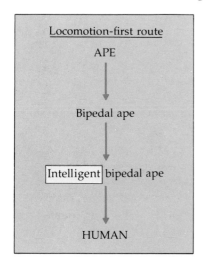

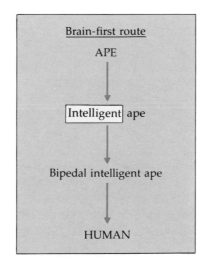

A scheme showing early ideas on human origins: brain first or last?

25

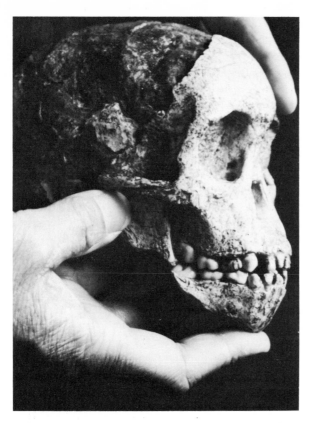

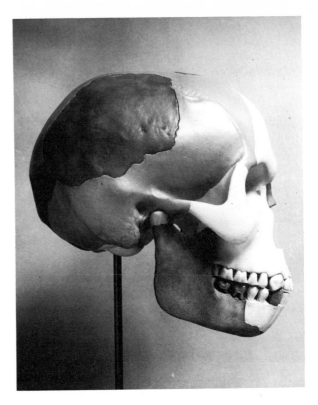

The Taung child. When Raymond Dart found this infant's skull in 1924 he recognized its hominid status and named it *Australopithecus africanus*; it was the first early hominid to be discovered. The British establishment ignored this skull for decades, arguing that it was the skull of an ape. Courtesy of Peter Kain and Richard Leakey.

Cast of Piltdown reconstruction, based on lower jaw, canine tooth and skull fragments (shaded dark). The ready acceptance of the Piltdown forgery—a chimera of a modern human cranium and the jaw of an orangutan—derived from the British establishment's adherence to the brain-first route. Courtesy of American Museum of Natural History.

ments have been met with considerable approval among palaeoanthropologists. 'It helps us keep the history of evolutionary theory in perspective', comments Sherwood Washburn, *emeritus professor* of anthropology at the University of California at Berkeley.

As mentioned in an earlier section, the two principle features of human evolution requiring explanation are the adoption of bipedalism and the expansion of the brain. Before the discovery of too many early hominid fossils, constrained speculation to any important degree, there were two major classes of theory. One held that the expanded brain came first; the other, that bipedalism was the precursor to a rise in intelligence. In both classes, it was often difficult to disentangle cause from effect.

The Piltdown forgery, a chimera of a relatively modern human skull and an orangutan's jaw, was, incidentally, a product of the 'brain-first' era. And when Raymond Dart found the first *Australopithecus*

africanus in South Africa in 1924, its hominid status was dismissed by the British palaeoanthropological establishment because it had a small brain. It belonged, according to Sir Arthur Keith, 'to the same group or subfamily as the chimpanzee and gorilla'.

Homo sapiens and its direct ancestors have diminutive canine teeth compared with those of most primates. As primates often employ their long sharp canines in displays of aggression, there has, therefore, been a great preoccupation until quite recently with how the early hominids defended themselves. In discussing the advantages of bipedalism for freeing the hands for manipulative skills, Darwin also noted that those same hands could deploy stones and clubs for killing prey and for defence against predators. Tools as weapons has been something of a persistent theme in stories of human origins; a line of argument that rose to its apogee in the 1960s with the graphic writings of Robert Ardrey. Much of the evidence adduced to support this argument—that many hominid fossils showed signs of intraspecific violence, and the supposedly irrepressible animal aggression in human nature—has now been shown to be wrong. It was surely no coincidence that the notion that humans were innately aggressive, and therefore that war was inevitable, arose at a time of rising concern about global hostilities.

In earlier times, there had been great interest in the idea that the fashioning and using of tools had been the prime engine of evolutionary change in the human lineage and that this had generated tremendous intellectual expansion. 'Man the Toolmaker' became a catch-phrase to describe the roots of humanity in the 1950s. This, coincidentally, was a period of great technological expansion.

The evolution of language as a prime focus of human history was fashionable as an idea a decade and a half ago. This, remember, was the age of Marshall Mcluhan, communications and the medium is the message. And it cannot be too surprising that following the distinctly macho image of 'Man the Hunter', popular in the 1960s and early 1970s, there came a counter-argument that the hominid line set out on its new evolutionary path with a female-focused social and economic unit. Gathering plant foods, not hunting red meat, was the prime economic activity, with the mother–infant bond holding the social unit together. Thus, following the hunting hypothesis, there came the gathering hypothesis, at a time when the women's movement was asserting itself.

The currently most popular notion combines elements of the previous two and can be described as the food-sharing hypothesis. The collection of plant foods provides the staple diet, with occasional lucky supplements of scavenged meat. This mixed economic activity is pursued from temporary home bases, the focus of complex social relationships. Because it is the currently most-favoured notion,

A discussion on the Piltdown Skull. From left to right: Mr T.O. Barlow, Professor Elliott Smith, Professor A.S. Underwood, Professor A. Kieth, Mr Charles Dawson, Dr Smith Woodward, Mr W.P. Pycraft and Sir E. Ray Lancaster. Courtesy of American Museum of Natural History.

– 1850s	Tools and defensive weapons	
– 1900s		
	Man the thinker	
– 1950s	Man the toolmaker	
	Man the communicator	
– 1960s		
	Man the hunter	
– 1970s	Man—woman, the gatherer	
– Present	Man/woman: the food sharers	

Changing fashions on the key influence in human evolution.

and, because it is based on relatively plentiful modern evidence, it is of course reckoned to be the truth; if not, then at least something close to it. History demands, however, that we be cautious.

9 / Modern Focus

One of the most striking features of palaeo-anthropology these days is the multidisciplinary nature of the pursuit. No longer is it a simple alliance between fossil hunters and archaeologists. Instead, a battery of sciences is being focused on the questions of human origins: in addition to geology, which provides the basic backdrop for the recovery of fossils, there is palaeoecology, taphonomy (the study of the way bones become buried), primatology, molecular biology, neurophysiology, energetics, and many more.

Instead of pondering the imponderable, such as the causal link between bipedalism, tool-making and use, and expansion of brain size, there is now a growing tendency to ask specific, answerable questions. What are the energetic considerations of upright-walking as opposed to, say, knuckle-walking? Under what ecological circumstances might it be energetically possible to evolve a large brain? What can the surface of fossil teeth tell us about ancient diets? What does one need to know about an assemblage of fossil bones and stone artifacts to be sure that the association is not mere coincidence? What does the history of animals contemporary with our ancestors tell us about, say, important migrations of the past? And so on.

The question of 'why we stood upright' is of course an old one and has engendered responses such as, 'so as to gain more visibility over tall grasses' and, 'so as to free the hands for defensive purposes, because we lost our long, sharp canines'. Neither scenario is very testable. But to ask, 'What are the energetics of bipedalism as compared to quadrupedalism?' as Richard Taylor of Harvard University has is a more promising approach. Other researchers have used Taylor's data and have added observations of their own.

For example, Peter Rodman and Henry McHenry, of the University of California at Davis, and Richard Wrangham, at the University of Michigan, examined the feeding habits of orangutans, which forage mainly for fruit in trees, and chimpanzees, which do the same but usually from sources that are more meagre and more widely dispersed. McHenry and his colleages wondered whether an ape, faced with food sources dispersed even more thinly than those exploited by chimpanzees, would forage more efficiently in energetic terms by knuckle-walking or by bipedalism. It turns out, although bipedalism is not particularly efficient compared with fully developed quadrupedalism in animals moving at high speeds, under the circumstances in which our ancestors evolved, that is, in the forest fringe with an ape-like ancestor, the evolution of upright walking was a very suitable adaptation. At slow walking speeds, bipedalism compares very favourably with quadrupedalism.

It may be, therefore, that bipedalism arose as an energetic strategy for foraging for widespread food sources. The point here is that the approach adopted by Taylor and others allowed specific questions, predictions and answers, a truly scientific way of tackling the problem.

Another exploration in energetics comes from the work of Robert Martin, at University College, London. Addressing primates in general and humans in particular, he asked the questions, 'What dietary and ecological factors affect a species' energy budget and reproductive strategy?' and, 'How is this related to brain size?' Some strands of the argument are as follows. Animals that adopt the reproductive strategy of having just a few carefully nurtured offspring, as against many that are left to fend for themselves, generally live in environments blessed with a stable food supply. And species that are endowed with large brains are those that feed on high energy diets. How does the requirement for a stable environment and a high energy diet match up to other conclusions about the environment of early hominids? Such questions are, thus, brought into scientific focus.

Parental care is common in the higher primates, and is especially pronounced in humans. Another strand of Martin's argument touches on the question of when extra parental care might have arisen in human evolution. Human babies are born with brains of about 350 cm^3 in volume, presumably the maximum size that human pelvic engineering can cope with. If human infants merely doubled their brain size as all other primates do, instead of quadrupling it as we in fact do, they would finish up with adult crania of about 700 cm^3 in volume. This figure happens to coincide with that for *Homo habilis*, the first hominid to show substantial increase in brain size in the hominid lineage. This result may be pure coincidence, or it may be a real insight into a major

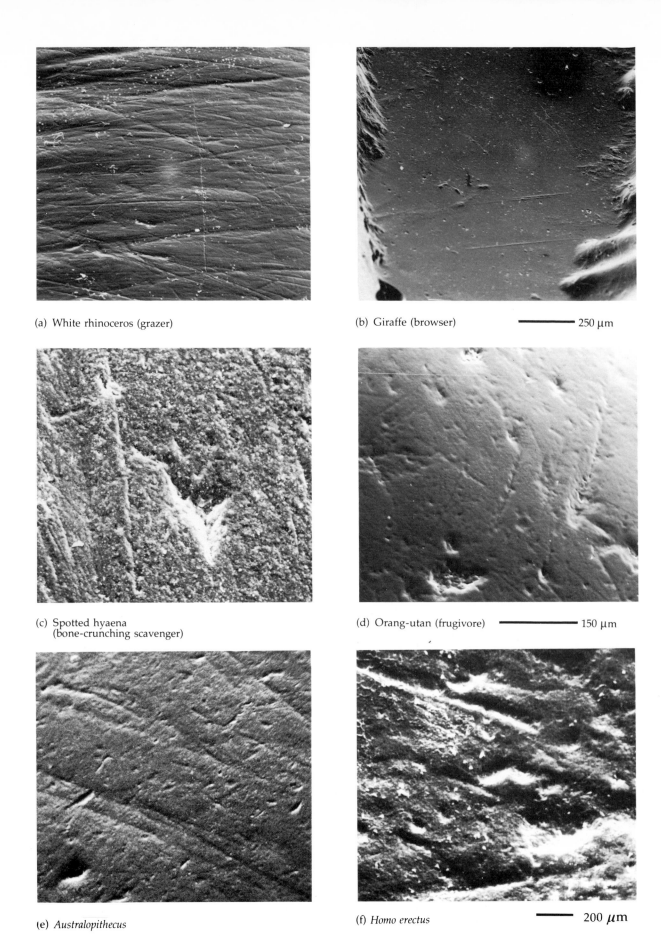

(a) White rhinoceros (grazer)

(b) Giraffe (browser) ━━━━━ 250 μm

(c) Spotted hyaena
(bone-crunching scavenger)

(d) Orang-utan (frugivore) ━━━━━ 150 μm

(e) *Australopithecus*

(f) *Homo erectus* ━━━━━ 200 μm

Speed	Walker	Energy use as a percentage of that predicted for a conventional quadrupedal animal
2.9 km/h	Chimpanzee	149 per cent
	Human	86 per cent
4.5 km/h	Chimpanzee	148 per cent
	Human	94 per cent
CONCLUSION: Human bipedal locomotion is as efficient energetically as conventional quadrupedalism; the chimpanzee's style of locomotion uses 50 per cent more energy than conventional quadrupedalism.		

Comparative energy efficiencies of humans (bipedal) and chimpanzees (quadrupedal) at slow walking speeds.

step in human evolution. Again, the point here is that an interesting question has been asked from an unusual perspective.

There has always been much speculation on the putative appearance of the last common ancestor between human and apes, but only infrequently has it been inspired by rigorous scientific questioning. By applying the principles of allometry in the study of body and limb proportions in monkeys, apes and humans, Leslie Aiello, again, of University College, London, has been able to dismiss some of the long-favoured 'models' for the human–ape ancestor because, it turned out, they were biologically incongruent. No, the last common ancestor was not like a chimpanzee; nor was it especially long in the forelimb, an adaptation to an arm-swinging life style; and neither was it like a modern Old World monkey. The comparative study tells us, says Aiello, that the last common ancestor probably moved about in a slow suspensory climbing motion, reminiscent of that of the modern howler monkey, ironically a New World monkey. This is not to suggest an involvement of New World primates in human origins; however it gives some guidance in interpreting the fossils of any putative last common ancestor.

This new approach of asking specific questions often brings unexpected answers, as happened to Alan Walker of Johns Hopkins University, Baltimore. When he imaged under the electron microscope the tooth surface of fossil hominids and compared them with marks on modern teeth he found that the early hominids and *Homo habilis* could be classified with chimpanzees, that is they were fruit-eaters. With *Homo erectus* came an apparently marked change in diet, possibly with the inclusion of underground tubers, possibly with increased meat-eating. This approach, still being developed, brings modern researchers as close as they will ever come to the food actually eaten by the individual whose fossilized remains are in their study. Walker had asked, 'What do the surfaces of chimpanzees' teeth look like?' 'What do baboons' teeth look like?' 'What do pigs' teeth look like?' 'What do hyaenas' teeth look like?' And, 'How does this help us understand better the diet of our ancestors?' He got an answer.

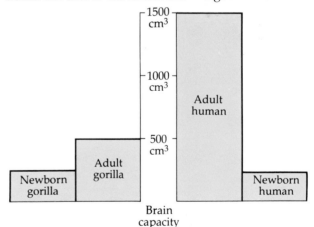

A scheme showing relative brain capacities in infant and adult humans and gorillas. Greater postnatal brain growth in humans increased selection for a more stable social organization.

Toothwear patterns under the electron microscope (see facing page). The presence of tiny silica crystals in grasses causes distinct scratches in the tooth enamel of grazers. Leaves on bushes and trees have no hard silica crystals and therefore polish the enamel of browsers to a smooth surface. Fruit-eaters encounter occasional small, hard objects in their diet that mark an otherwise smoothed enamel surface. Bone-crunching carnivores inflict deep scratches and pits in their enamel.

All the early hominids, and earlier hominoids such as *Sivapithecus* and *Ramapithecus*, have toothwear patterns most like those of frugivores (*Australopithecus robustus* is shown here). Only with the arrival of *Homo erectus* 1.6 million years ago does the pattern change; this animal was a meat-eater that either crunched bone (which is unlikely) or included hard grit in its diet, perhaps attached to underground tubers. Courtesy of Alan Walker.

10 / Science of Burial

The fossil and archaeological records are the principal sources of evidence upon which human prehistory is reconstructed. Unless that evidence can be interpreted with some confidence, the reconstruction, however convincing, may not be valid. During recent years there has developed a tremendous emphasis on understanding the multifarious processes that impinge on bones and stone artifacts that become part of the record. The science of taphonomy, as this pursuit is known, has revealed that the prehistoric record is littered with snares and traps for the unwary.

A taphonomist in a pessimistic mood has been heard to argue that, because of the countless complicating factors that can plant false clues in the record, the chances of reconstructing the past are virtually nil. There is more generally, however, a sense of optimism that step by step specific problems in taphonomy are being solved. Through a combination of ever more careful study of material from the prehistoric record and the development of ingenious experiments and observations on modern material, it is becoming possible to scrutinize the material evidence of human history with the required degree of confidence.

Death is a bewildering, dynamic process in the wild. First, many animals meet their end in the jaws of a predator rather than passing away peacefully in their sleep. Once the primary predator has had its fill, scavengers, which in modern Africa would be hyaenas, jackals, vultures and the like, move in. The carcass is soon stripped of meat and flesh and the softer parts of the skeleton, such as vertebrae and digits, are crushed between powerful jaws. The remaining bones dry rapidly under the sun. Even in this initial phase the skeleton is probably partially disarticulated, hyaenas having torn off limbs and other body parts to be consumed in the crepuscular peace of their dens. Passing herds of grazing animals bring a new phase of disarticulation and disintegration as hundreds of hoofs kick and crush the increasingly fragile bones. Within a few months of a kill the remains of, for example, a zebra, will be scattered over an area of several hundred square metres, and a large proportion of the skeleton will

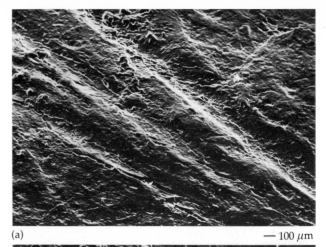

(a)   — 100 μm

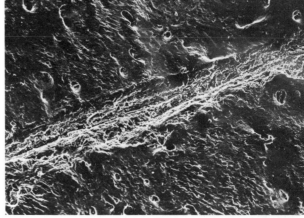

(b)   — 100 μm

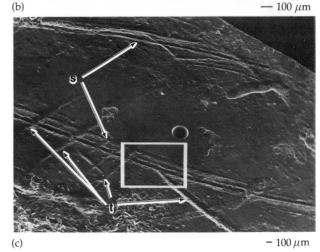

(c)   — 100 μm

Bone surfaces under the electron microscope.
A. The surface shows the round-bottomed groove made by a hyaena gnawing at a modern bone.
B. A sharp stone flake makes a V-shaped groove in a bone surface (modern).
C. This fossil bone from the Olduvai Gorge carries carnivore tooth marks (t) and stone flake grooves (left at centre): the scavenger activity followed the hominid's on this occasion. Courtesy of Pat Shipman and Richard Potts.

apparently be missing. Some of the skeleton may indeed be miles away, mouldering in a hyaena's den. Some bones will have been shattered and disintegrated into minuscule pieces. Other bones will have been compressed into the ground by the pressure of passing hoofs. Only the toughest skeletal parts, such as the lower jaw and the teeth, remain intact.

As such fate awaits most animals in the wild, it is perhaps not surprising that the fanfared announcements of ancient hominid discoveries typically mean an interesting jaw, or arm bone, or, rarely, a complete cranium has been found. The most complete skeleton of an early hominid unearthed so far is the famous 'Lucy', whose collection of fragmented three million-year-old bones represents just 40 per cent of her original self. Nothing else approaches this degree of completeness.

In order to become fossilized a bone must first be buried, preferably in fine alkaline deposits and preferably soon after death. Most hominid fossils have been found near to ancient lakes and rivers, partly because our ancestors, like most mammals,

were highly dependent on water; and partly because these provide the best depositional environments where fossil formation is favoured.

As it happens, the forces that can bury a bone—for example, layers of silt from a gently flooding river—can later unearth it as the river 'migrates' back and forth across the flood plain through many thousands of years. When this occurs the bones are subject once again to sorting forces: light bones will be transported some distance by the river, perhaps to be dumped where flow is slowed, while heavier bones are shifted only short distances. Anna K. Behrensmeyer, a leading taphonomist, identifies transport and sorting by moving water as one of the most important taphonomic influences. Abrasions caused when a bone rolls along the bottom of a river or stream are tell-tale signs of such activity, as are the characteristic size profiles and accumulations in slow velocity areas of an ancient channel. For hominid remains, the result of all this activity is often the accumulation of hundreds of teeth, and little else, as the researchers working along the lower Omo River

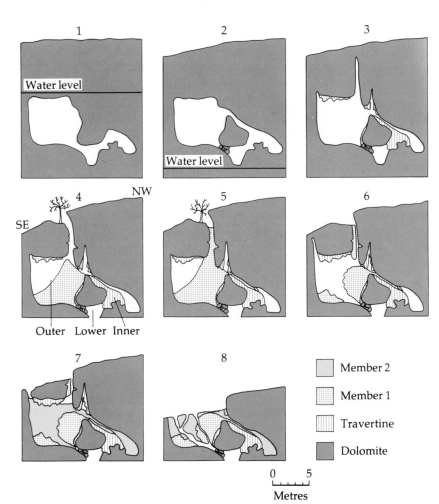

Formation of the Swartkrans Cave, South Africa. An irregularly shaped cave was dissolved out of the dolomite; when the water level dropped a chimney opened to the outside along a weakness in the dolomite, and debris, including bones, entered, eventually to form the breccia of member 1 (1–4); debris choked the chimney (5), which remained blocked for perhaps one million years; further erosion reopened the chimney (6); member 2 breccia formed (7), all of which was exposed as the hillside eroded away (8). A very large collection of *Australopithecus robustus* and some *Homo erectus* specimens has been recovered from this complex site. Courtesy of C.K. Brain.

in Ethiopia know only too well.

Large numbers of hominid fossils have been recovered from the rock-hard breccia of a number of important caves in South Africa. At one time, it was thought that the hominids lived in the caves and that the bones of other animals found with them were remains of food brought there to be consumed in safety. In addition, the fractures and holes present in virtually all the hominid remains were considered to be the outcome of hominid setting upon hominid with violent intent. In many ways, the South African caves represent one of the most severe taphonomic problems possible, but with years of patience a group of workers (and in particular C.K. Brain) has cut through the first impressions and progressed a little closer to the truth.

Most of the bone assemblages were almost certainly the remains of carnivore meals accumulated over very long periods of time. The profile of skeletal parts present matches what would be expected after carnivores had eaten the softer parts. And the damage recorded in the hominid crania was simply the result of rocks and bones compressed into them as the cave deposits mounted. Exactly how much time is represented in these fascinating accumulations is difficult to determine. But the question, as in many taphonomic investigations, is a key one.

The earliest putative living camp site, dated at about two million years, is at Olduvai Gorge; the frequency of such sites increases through the archaeological record. The question of determining whether or not a collection of animal bones and stone fragments represents the focus of hominid activity is more difficult than might be expected. Part of this problem, as Pat Shipman of Johns Hopkins University stresses, is the obvious—but usually overlooked—fact that these creatures were not humans and therefore did non-human things that might be difficult for us at this distance to recognize. Electron microscopic analysis of fossil bone surfaces can distinguish between marks made by carnivore gnawing and by stone flakes used to deflesh the bone. Indeed, it is now possible to search for patterns in bone accumulations and breakage characteristics that tell of a practised hunter.

According to Shipman, it is only with the appearance of *Homo erectus* some 1.5 million years ago that the taphonomic patterns in the bones and stones begin to suggest that hunting was becoming an important economic activity. Could it be mere chance that this coincides with a decrease in the frequency with which hominid fossils are found at any particular time horizon? Perhaps not. Carnivores require a much larger territory and, therefore, have lower population densities than plant-eaters.

11 / Hominid Precursors

The Miocene epoch (25 to 5.5 million years ago) saw the beginnings of the modern mammalian faunas. It was also a time of tremendous adaptive radiations among the higher primates of the Old World, first among the apes and then among the monkeys. As the end of the epoch approached, the fortunes of the apes dwindled while those of the monkeys rose, events that in some degree at least were probably causally related. The end of the epoch also witnessed the emergence of an ape stock that gave rise to the African apes on the one hand and the hominids on the other.

Ideas about the origin of the hominid lineage have undergone something of a revolution in the past few years. This has been due, in part, to evidence from molecular biology and to the discovery in 1980 of a particularly fine specimen of a creature known as *Sivapithecus indicus* from the foothills of the Himalayas in Pakistan. The results of the revolution have been twofold. First, *Ramapithecus*, a close relative of *Sivapithecus* that had long been a favourite candidate as the first hominid, was shifted off the main path of human ancestry. Second, the date of the divergence between humans and the African apes is considered to be later than the previous consensus figure of 15 million years ago and is now put at somewhere between ten and five million years ago. There are, of course, dissenting voices on both counts, but these are now in a minority.

Agreement on what is *not* a human hardly seems like grand progress, and in isolation the relegation of *Ramapithecus* from hominid status would be less than constructive. Overall, however, a coherent story is beginning to emerge, even if the principal

EARLIER VIEW   Present   **CURRENT VIEW**

Changing ideas on hominoid ancestry. Earlier views on hominoid ancestry envisaged ladder-like progressions, with *Ramapithecus* branching off as the first hominid at least 15 million years ago. A large gap was placed between the hominids and the apes. Hominoid history, like the history of most animal groups, is now seen as a much more intricate bush-like structure. *Ramapithecus* is no longer thought of as the first hominid, and the gap between hominids and apes, particularly the African apes, has been closed.

35

character, the last common ancestor between African apes and humans, has yet to put in an appearance. Like all stories, this one is best told from the beginning.

The earliest fossil evidence of an ancient apestock comes from 28 million-year-old deposits in the Fayum Depression in Egypt, an arid area that at the time was lush green with tropical forests, woodland and winding rivers. The principal specimen from there, a cat-sized creature known as *Aegyptopithecus zeuxis,* although not an ape itself, probably gave rise to apes of the later Miocene, according to its discoverer Elwyn Simons, of Duke University.

A putative descendant of the Fayum anthropoids, the 18 million-year-old *Proconsul africanus* from Kenya, is an important fossil clue to the past, made even more so by the recent rediscovery in a Nairobi museum collection of many bones that represent almost 75 per cent of a skeleton. With so much evidence to call on (a rare thing in this business), it has been possible to infer with confidence something of locomotor characteristics of the body and its relation to the cranium above it.

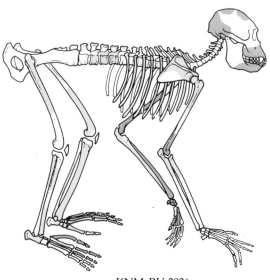

KNM-RU 2036

A reconstruction of *Proconsul africanus* based on material found prior to 1959 (in peach) by Mary Leakey and in 1980, among the Nairobi Museum collections, by Alan Walker and Martin Pickford. The individual, which was a curious mixture of monkey and ape characteristics, was a young female who lived about 18 million years ago. Courtesy of Alan Walker.

Proconsul is a curious mixture of monkey and ape characteristics. Its relatively long trunk is monkey-like, so are its arm and hand bones; its shoulder and elbow regions are ape-like. Strikingly ape-like, too, is the head, which is large in relation to the diminutime 11 kg body, and the dentition, which includes small molar teeth and large projecting canines. *Proconsul* must have run quadrupedally along branches, as modern monkeys do, and apparently ate relatively soft fruit, the typical ancestral diet of the hominoids.

Around 18 million years ago Africa rejoined with Eurasia and there followed a rich exchange of faunas. Monkeys and apes moved into Eurasia, and the apes radiated wonderfully. Outnumbering monkey species by as much as 20 to 1, apes became a very diverse group, though their principal habitat was forest and their principal diet frugivorous. There were several major groups of Miocene apes, but the one of greatest interest here includes *Ramapithecus, Sivapithecus* and an impressively large relative known as *Gigantopithecus.* Together, they may be described as ramamorphs—a term introduced into

Photographic comparison of *Gigantopithecus* and *Homo erectus.* The partially-reconstructed, one million year old *Gigantopithecus* mandible (left) from China dwarfs the *Homo erectus* jaw. The enormous size of *Gigantopithecus* must have made it a formidable ground-living ape. Courtesy of Peter Kain and Richard Leakey.

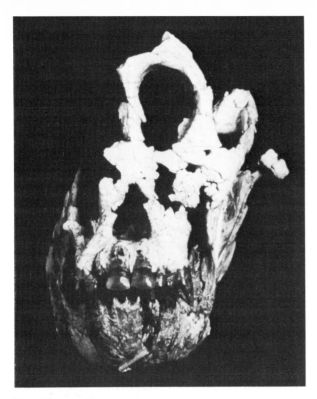

Sivapithecus. This eight million year old specimen of *Sivapithecus indicus* comes from the Potwar Plateau in Pakistan. The animal was about the size of a chimpanzee but had the facial morphology of an orangutan; it ate soft fruit (detected in the toothwear pattern) and was probably mainly arboreal. Courtesy of David Pilbeam.

the literature by David Pilbeam—most species of which lived between 18 and seven million years ago.

Ramamorphs are altogether more ape-like than *Proconsul,* and, as such, their method of locomotion was likely to be beneath-branch climbing, using longer limbs on a shorter body. One set of features that for a long time fixed the attention of palaeo-anthropologists, however, was the jaws and teeth, partly because most of the fossil collections consisted mainly of these parts but, more significantly, because of their resemblance to later known hominids. Rama-morphs had large cheek teeth covered with thick enamel that were set in robust jaws; the canines were less prominent than in true apes; and compared with the incisors of *Proconsul,* those in the ramamorphs appear more suited to crushing food items.

As these dental features, particularly the thick enamel on large molar teeth, were assumed to be specialized characters of the hominids, the natural conclusion was that the earlier group gave rise to the later one. *Ramapithecus,* as the smallest of the ramamorphs, weighing perhaps 18 kg, was the best candidate as the direct ancestor to *Australopithecus,* and was therefore designated as the first hominid. As *Ramapithecus* existed 14 million years ago, the split between the hominids and African apes was deduced to be at least this ancient.

Although doubts about this early divergence date had begun to concern many workers, principally from the molecular evidence, it was fossil evidence that finally caused the revolution. Early in 1982, Pilbeam published details of a *Sivapithecus* face found in 1980, which, he argues, had strong anatomical affinities with the modern orangutan. In June of that year, Peter Andrews of the Natural History Museum, London, and Jack Cronin of Harvard, published a paper in which they essentially concurred with Pilbeam. If certain specialized an-atomical characters align *Sivapithecus* with the orangutan, then the other ramamorphs must be placed in this broad ancestral group too. *Ramapithe-cus,* the argument runs, therefore cannot have been a hominid, since it belonged to a specialized group (leading to modern orangutans) that split off from the common ancestral stock.

In any case, the thick enamel on large molars is now regarded as a primitive—not a specialized—character that was present in the ancestor common to ramamorphs and the stock leading to African apes and humans. This trait continued in the rama-morphs and the early hominids, but the African apes became specialized with thin enamel and with other characters too.

The discovery in late 1983 of an 18 million year old *Sivapithecus* species from northern Kenya has led Alan Walker and Richard Leakey to suggest that the ramamorphs might, in fact, represent the basic hominoid stock of the time—that is, the common ancestral group to later apes and hominids—and not to have diverged along a specialized route towards the modern orangutan, as proposed by Pilbeam and Andrews. In which case, the orangutan would be considered as something of a living fossil, showing some of the primitive features of the earlier hominoid stock. African apes would then be seen as being evolutionarily specialized in a number of respects. The matter is still up for debate.

12 / The First Hominids: the Fossils

The earliest undisputed hominid remains come from two separate East African sites, each remarkable in its own way. One site, in the Hadar region of Ethiopia, has yielded several hundred fossil fragments, including the partial skeleton known as Lucy, of individuals that lived and died near a now-vanished lake between 3.0 and 3.6 million years ago. The second site is called Laetoli, located 50 km south of the famous Olduvai Gorge in Tanzania, where three hominids left a 20 m trail of footprints in volcanic ash some 3.75 million years ago. In addition, fossil fragments of about 13 individuals—mainly teeth and jaws, together with a few postcranial bones—have been found in these same sediments.

The one unequivocal conclusion to be drawn from these important finds is that by at least 3.75 million years ago upright walking had already developed to an advanced degree while brain size remained very modest, only slightly in excess of 400 cm. There are indications from preliminary finds at a site south of Hadar, the Middle Awash, that this specifically hominid characteristic was already in existence by four million years ago, a date that approaches the inferred divergence between hominids and the African apes according to certain molecular biological evidence.

A ready description of these earliest hominids is that they were built as from an ape's head on top of a human-like body. In fact, although the head has many primitive ape-like features, there are enough hominid characteristics apparent to place it firmly within the human lineage. Although these creatures undoubtedly were habitual bipeds when they were moving over the ground, there are enough primitive features to root them close to an ape-like ancestry and, incidentally, to indicate a considerable arboreal facility. Overall then, these earliest hominids were a chimera of ancient and modern throughout.

Donald Johanson, an American anthropologist who led the Hadar expeditions jointly with French palaeoanthropologists Maurice Taieb and Yves Coppens, has studied the Hadar remains intensively with Tim White, another American at the University of California at Berkeley. They note that the tooth structure in the jaw of modern apes is somewhat like three sides of a rectangle, with two prominent canine teeth marking the two corners. A modern human jaw, by contrast, is much more like an arch, and the canines (and incisors for that matter) are greatly reduced. A diastema, or gap, between the canine and the first incisor on each side is typical of apes but is rare both in modern humans and in hominids from two million years onwards. Hominid cheek teeth are

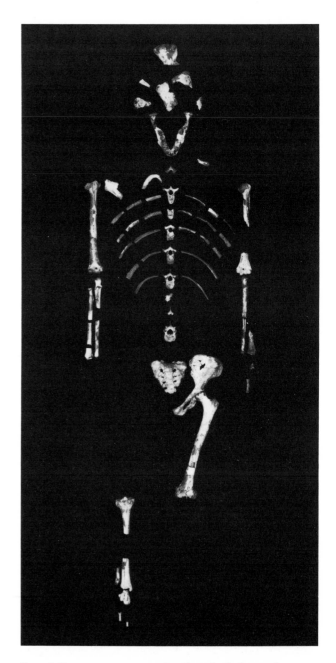

'Lucy'. Forty per cent complete female skeleton of *Australopithecus afarensis*. Courtesy of Cleveland Museum of Natural History.

rather large and flat and are capped with thick enamel.

When Johanson and White compared the dental arcade of the Hadar specimens with these well-established patterns they found these earliest hominids to be somewhat intermediate between apes and two million-year-old hominids in overall shape, in the size of the canines and in the prominence of the cusps on the cheek teeth. Moreover, about half of the Hadar hominids do have a gap between the canines and the incisors, but not as marked as in apes; and the first premolar is more typically ape-like than hominid-like, in that, in some individuals at least, it has one cusp rather than two. The large, flat cheek teeth are very similar to those seen in later hominids.

specimens as a new species of hominid, *Australopithecus afarensis,* a bold decision that continues to be contentious (see following section).

The postcranial skeleton of Lucy and her companions is human-like in many respects. Most striking is that the squat structure of the pelvis, as compared with the elongated form in apes, indicates some form of bipedality. The structure and angle of the head of the femur, the angle of the knee joint and the platform nature of the footbones are also consistent with bipedality. The opposable great toe, characteristic of modern apes, is lost in these hominids: the great toe is locked in line with the lateral toes, as it is in humans.

Although Lucy's pelvis is most definitely not that of an ape, neither is it fully human in form, par-

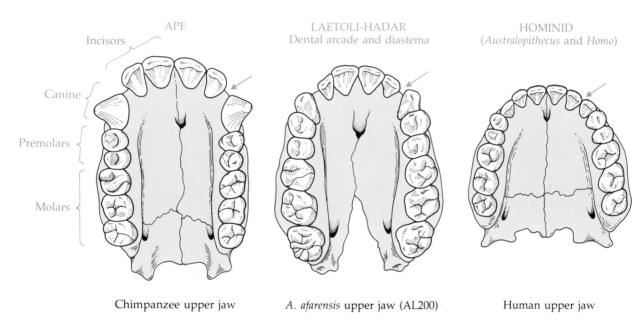

APE

Incisors

Canine

Premolars

Molars

Chimpanzee upper jaw

LAETOLI-HADAR
Dental arcade and diastema

A. afarensis upper jaw (AL200)

HOMINID
(*Australopithecus* and *Homo*)

Human upper jaw

Comparison of dentition in the palates of an ape, human and *Australopithecus afarensis. A. afarensis* dentition is a mixture of hominid and ape characteristics. The incisors are relatively large, like an ape's; and in 45 per cent of specimens there is a gap, a diastema, between the canine and incisor. Such a gap is uncommon in later hominids. The canines are not as large as in apes, but neither are they as small as in some of the later hominids. The premolars are more

primitive than those of later hominids, but the molars are very hominid-like in being large and usually showing a lot of wear to the cusps that generally makes the teeth rather flat. The cheek teeth in this specimen are in a straight line, like those of an ape, except for the last molar, which turns in a little to give a curve to the row. Hominid tooth rows from later taxa are generally arched as shown. Courtesy of Luba Gudtz.

This list of characters intermediate between those of apes and the previously known earliest hominids (from around two million years ago), and similar inference from other parts of the cranium and skeleton, persuaded Johanson and White that the Hadar (and Laetoli) specimens represented a new species of hominid that was ancestral to all later hominids. In 1978, therefore, they designated these

ticularly in the angle of the iliac blades. Nevertheless, concludes Owen Lovejoy of Kent State University, biomechanical and anatomical studies of the mosaic pelvis indicate that the structure is consistent with a style of bipedality that is strikingly modern. By contrast, two researchers at the State University of New York at Stony Brook interpret the mixture of characters in Lucy's pelvis as indicative of a some-

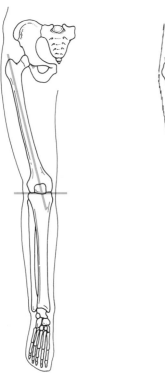
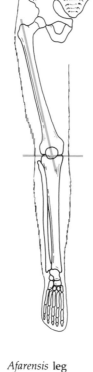
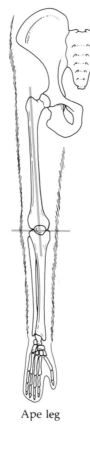

Diagram showing different valgus angles in humans, apes and *Australopithecus afarensis*. The angle subtended by the femur at the knee, the valgus angle, is critical to bipedal locomotion. With the femur angled as in humans, the foot can be placed underneath the centre of gravity while striding. An ape's femur is not angled in this way, and so it 'waddles' during bipedal locomotion. The valgus angle of *Australopithecus afarensis* is human-like, indicating its commitment to bipedality. Note also, the human-like shape of the *afarensis* pelvis. Courtesy of Luba Gudtz.

Human leg   *Afarensis* leg   Ape leg

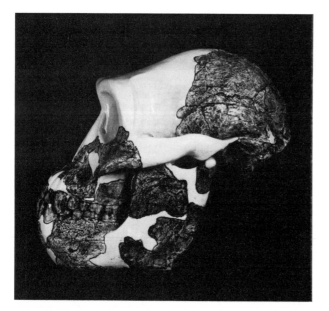

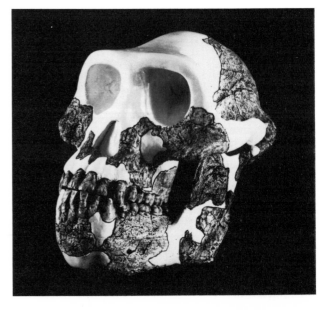

Lateral (left) and three-quarters facial view (right) of skull reconstruction representing *Australopithecus afarensis*.

Courtesy of Cleveland Museum of Natural History.

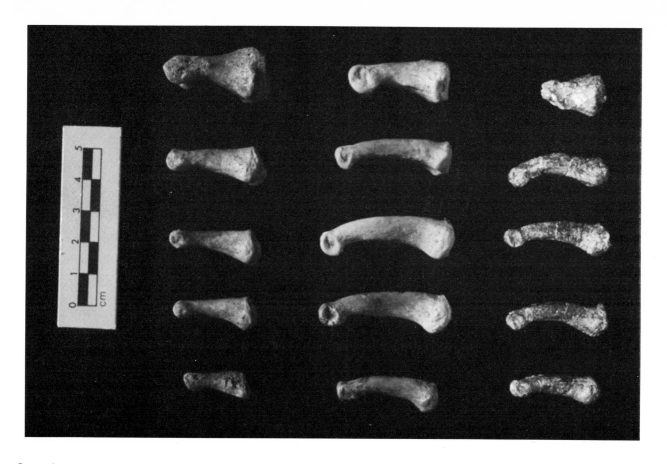

Lateral view of the Hadar proximal phalanges (toe bones) (right column) compared with those of chimpanzee (centre) and human (left). Courtesy of Bruce Latimer, Cleveland Museum of Natural History.

what simian form of bipedality, a bent-hip, bent-knee gait. This difference of opinion is yet to be resolved.

Studies on the Lucy skeleton and on other Hadar specimens show *A. afarensis* to have had long fore-limbs and relatively short hindlimbs—an ape-like configuration. (Milford Wolpoff, of the University of Michigan, argues, however, that Lucy's small legs are the length one would expect in a human of her diminutive stature.) Even more ape-like are the distinctly curved finger and toe bones. The Stony Brook researchers, Randall Susman and Jack Stern, interpret these features as adaptations to significant arboreality. Others, including Lovejoy and White, suggest other interpretations might be possible.

One important feature of the Hadar hominids is the great range in body size implied by the fossils, ranging from Lucy who probably stood a little over 1 m tall and weighed 25.0 kg to large individuals standing 1.7 m tall and weighing perhaps 52 kg. Johanson and White have interpreted these differences as differences between the sexes (but see next section). Such sexual dimorphism is seen in some primate species, such as gorillas, in which males are much bulkier than females.

Extreme sexual dimorphism in body size is typically associated with a polygamous social structure, with one male maintaining a harem of several females. The large body size of the male is important, in part at least, in competition with other males. Extreme dimorphism in canine size is usually associated with this social arrangement too, but this dimorphism is very much reduced in *A. afarensis*. Although the canines in this hominid are large compared with those of later hominids, they are much reduced compared with those of an ape.

13 / The First Hominids: the Theories

When in 1978 Donald Johanson and Tim White named the Hadar and Laetoli hominids *Australopithecus afarensis* it was indeed a bold step: no new hominid species had been named for almost 15 years. Furthermore, carrying their deductions to logical conclusions, they published a landmark paper in *Science* in January 1979 in which they presented a new phylogenetic tree describing human origins: *Australopithecus afarensis* was placed as the ancestor of all subsequent hominids.

Johanson and White's new phylogeny quickly gained a good deal of support in the palaeoanthropological community but it also attracted its critics. Even seven years after the publication of the *Science* paper, when many more workers have had an opportunity to examine the Hadar fossils or research casts, the subject of relationships in the early stages of the human family tree still remains highly contentious.

One key issue emerged, upon which the acceptance or rejection of the Johanson/White hypothesis depends crucially. This issue concerns the number of hominid species that are represented by the Hadar and Laetoli specimens. Is there just one species, *Australopithecus afarensis,* as Johanson, White and numerous other workers argue? Or, Are there two species, or perhaps even more, as suggested by Kenyan palaeoanthropologist Richard Leakey, French palaeoanthropologist Yves Coppens, who was a member of the team that found the Hadar specimens, and numerous others? Clearly, if more than one species of hominid lived in East Africa between three and four million years ago, the fossils known as *Australopithecus afarensis* cannot be ancestral to all other hominids.

The remarkable selection of fossils that have been retrieved from the Hadar deposits is striking, among other things, for the difference in size between the smallest and the largest individuals, a difference reflected not only in the skeletal parts but also in the canine teeth. Johanson and White attribute these size differences to sexual dimorphism, a characteristic common to primates. Others, such as Alan Walker and Richard Leakey, have contended that in relative terms the dimorphism is greater in *afarensis* because overall it was a smaller animal than the gorilla. If the size dimorphism did indeed exceed modern examples, this would not necessarily be fatal to the proposition that there was only one species at the Hadar, but it would demand caution in the interpretation of the remains.

The much smaller selection of fossils from Laetoli, collected by Mary Leakey, are of course formally included in the species *Australopithecus afarensis.* It is noteworthy that the size range here is not as marked as among the Hadar fossils. Does this mean that the Laetoli material cannot after all be lumped with the Hadar fossils and that several species did indeed exist at this time period? Mary Leakey argues so, and she suggests that the Laetoli specimens are in fact an early form of *Homo,* not *Australopithecus.*

As mentioned in the previous section, analysis of locomotor adaptations has, to some eyes at least, revealed considerably more anatomical features suited to arboreality in the smaller individuals than in the larger ones. Beyond noting that this putative sex difference is larger than that observed in any living ape, Stern and Susman, who did much of the work on locomotor styles, have made no comment about the numbers of species represented at the Hadar. Christine Tardieu, a colleague of Coppens at the Musée de l'Homme in Paris, has concluded, however, from her work on the knee structure from various Hadar specimens, that there are two distinct species.

Coppens, who coauthored the 1978 paper with Johanson and White, announcing the new species name of *Australopithecus afarensis,* has since changed his mind and is even more specific than Tardieu. One species is indeed an archaic *Australopithecus,* he says. However, a second species he sees as being much more modern, a primitive *Homo* perhaps. This conclusion, therefore, aligns Coppens' with Mary Leakey's position, and, incidentally, with Richard Leakey's too. There are differences in the dental architecture between the large and small individuals that suggest two separate species, says Richard Leakey; one of them is a primitive *Homo,* he argues. Very recently, there has emerged the suggestion—from several different sources and based on several different anatomical characteristics—that at least some of the Hadar hominids display specializations diagnostic of the robust *Australopithecus.*

Johanson and White have, of course, had the opportunity to spend much more time studying the

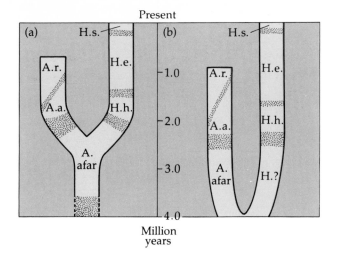

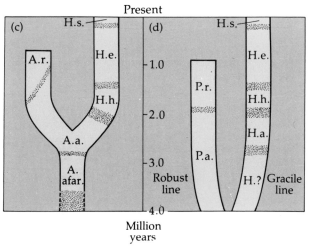

There is no generally agreed phylogenetic tree for describing human evolution. Although there are several proposed branching patterns, perhaps the major distinction between them is whether *Homo* is perceived as a late arrival (a and c) or as an early arrival (b and d). The second point of discussion is the relationship between *Australopithecus africanus* and *Australopithecus robustus/boisei*. Although many believe the one to be ancestral to the other (a, b and c), some consider the robust australopithecines to be quite separate from the earlier gracile species (d, in which *Paranthropus robustus* is equivalent to *A. robustus* and *Homo africanus* is equivalent to *A. africanus*). A third point to note is whether *Australopithecus africanus* is ancestral to *Homo* (c and d) or is a side issue (a and b).

Hadar material than anyone else; White has studied the Laetoli fossils too. Their response to the assertions that certain anatomical features indicate the presence of more than one species is to argue that the differences are merely the result of scaling differences between large and small individuals.

They also say that one group of Hadar fossils—a collection of more than 200 fragments from at least 13 individuals—appears to have resulted from the catastrophic death of a 'family' of hominids. As they were all together in one place at one time, they are likely to be of the same species. There is, however, some doubt about the real nature of the death of this group, and some taphonomists suggest that it is just as likely to be an accumulation of individuals over a period of tens or even hundreds of years, perhaps at the base of a tree used by carnivores, for example. This is not an argument *against* the single species proposition, but it means that the 'first family', as they have been dubbed, cannot be used as an argument *for* the proposition.

There are currently three main hominid phylogenetic trees, each with its own cadre of proponents, plus many minor variations. First, there is Johanson and White's, which places *Australopithecus afarensis* as ancestral to the later australopithecines on the one hand and the *Homo* line on the other. This gives the *Homo* line a relatively late start in human prehistory. By contrast, there is the view, supported by the Leakeys for example, that a common ancestor of the *Homo* lineage and the australopithecine lineage diverged quite early in history, and certainly by the time *Australopithecus afarensis* was on the scene. A third view is a combination of the two: *afarensis* is accepted as being ancestral to a gracile australopithecine, *Australopithecus africanus*; and it is proposed that *africanus* gives rise to the *Homo* line on the one hand and another australopithecine, *Australopithecus robustus*, on the other.

This diversity of opinions demonstrates very clearly one point: there are not, as yet, enough fossils recovered to provide a firm answer. However, the differences of professional opinion should be seen as lively and creative, not negative and destructive.

14/The Australopithecines

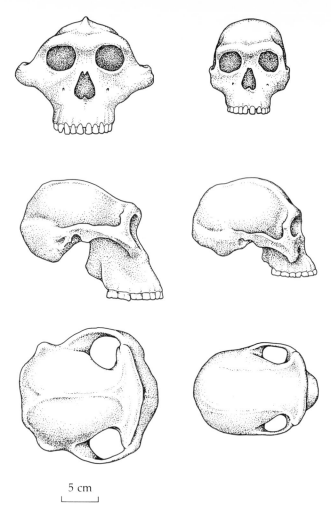

By two million years ago there were several well-established hominid lineages in Africa. One was the *Homo* line that eventually went on to give rise to modern humans. Another was the australopithecine line, which suffered the fate common throughout the history of life: extinction without issue.

Two main australopithecine species have been recognized. The first is *Australopithecus africanus*, a 1.3 m high, 40 kg creature, whose remains have been found in the breccia of South African caves and in ancient lake deposits in East Africa. (There is, incidentally, some discussion about the exact identity of the latter.) The second is an altogether stockier animal, *Australopithecus robustus*, which stood 1.75 m high and weighed 60 kg. Fossils of this have been recovered from both South and East Africa. There appears to have been some geographical variation in this species, however, as those individuals from more northerly regions are even more robust than their southerly cousins. In recognition of this difference, the East African robust australopithecines are generally referred to as *Australopithecus boisei*, that is, a different species.

Although, as indicated, there is some doubt about whether *Australopithecus africanus* is the ancestor of the *Homo* line, there is more general, although by no means universal, agreement that the gracile australopithecine gave rise to the more robust species. As far as can be judged—because of certain difficulties in dating the South African caves, for example—*Australopithecus africanus* arose somewhere between two and three million years ago, while the origin of the robust species is probably closer to two million years ago. *Robustus* apparently became extinct about one million years ago, which is somewhat later than the demise of the gracile species.

The first australopithecine to be discovered came from a limestone quarry near Kimberly, South Africa in 1924. Raymond Dart recognized the partial cranium, brain endocast, face and jaw of a young child as ape-like but definitely not an ape, a display of perspicacity for which he was initially rewarded with searing scorn. This specimen was named *Australopithecus africanus*, many more examples of

5 cm

Graphic reconstructions of the crania of the robust australopithecine, *Australopithecus boisei* (left), and the gracile australopithecine, *Australopithecus africanus*, from Koobi Fora in Kenya. The massive buttressing for muscle attachment in the larger species is particularly evident in the dorsal view (bottom). From *The Hominids of East Turkana* by Walker A. and Leakey R.E.F. Copyright © (1978) by Scientific American, Inc. All rights reserved.

which have subsequently been recovered from other caves in the area, specifically Sterkfontein and Makapansgaat.

Fourteen years after the first *Australopithecus africanus* was found, remains of a second, more robust hominid were found in another limestone cave, Kromdraai. Robert Broom, a fossil hunter of equal enthusiasm to Dart, named the new hominid *Paranthropus robustus*. Over the next two decades there was a flurry of discoveries and an almost equally enthusiastic naming of fossil hominid species. Eventually, there was a much-needed rationalization of names, giving the *Australopithecus africanus* and *robustus* with which we are now familiar. (There are specimens of early *Homo* from the South African caves too.) With the discovery of further cranial and postcranial material from the gentler deposits of East Africa, particularly from Olduvai Gorge in Tanzania and Koobi Fora in Kenya, a relatively clear palaeontological sketch can be drawn of these hominid species.

The australopithecines' brains were relatively small, measuring about 480 cm³ in the gracile species (in the range of a modern chimpanzee) and 550 cm³ in the larger, more robust species. The crania, therefore, were rather ape-like, but the face in both species was shorter than in apes, less protruding and substantially buttressed (particularly in the robust species) by a characteristic boney architecture. Both species are equipped with robust lower jaws into which are set rather large and flat molars and premolars, which appear to be specially adapted to

microwear patterns in the enamel indicates that both ate an essentially frugivorous diet, rather like that of a modern chimpanzee.

One feature that alerted Dart to the special nature of the Taungs specimen was the position of the foramen magnum, the aperture through which the spinal cord leaves the cranium; it is located much more towards the middle than in apes, indicating that the creature probably walked upright.

The evidence for bipedal posture in the australopithecines is strong. The vertebral column shows the typical hominid curvature and the pelvis is shorter than in apes, although it is not as short and as buttressed for weight bearing as in modern humans. The shape of the femur is clearly that of a bipedal animal, although there are interesting differences in a modern human thigh bone. The head of the femur, which forms the ball and socket joint with the pelvis,

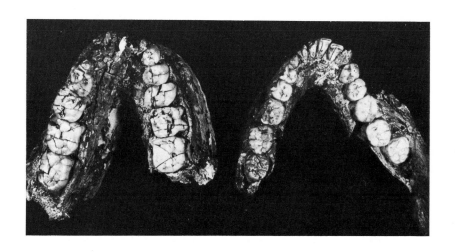

Comparison of lower jaws of *Australopithecus robustus* (left) and *A. africanus* (right). Note the massive molar teeth in the *Australopithecus robustus* mandible from Swartkrans (left) compared with that of *A. africanus* from Sterkfontein (right). Courtesy of Milford Wolpoff.

efficient grinding. The incisors form a row of vertical slicing teeth, which are joined without a gap by diminutive, rather flat canines. The molars in the robust species are extremely large, a feature that earned the first such fossil from Olduvai Gorge the sobriquet 'Nutcracker Man'. By contrast, the front teeth are very small indeed, often smaller than those of *africanus*.

The masticatory machinery of robustus was so massive that most individuals possessed a bony crest running along the midline of the skull to which the jaw muscles attached. Despite this, it appears that robustus was equipped to process merely a larger volume of food than *africanus* and not something that was substantially tougher. The pressures exerted on the thick-enamel covered tooth surface in the two species were very similar, and analysis of

is constructed in modern humans from a relatively large round ball on a short cylindrical arm. In the australopithecines, the ball is smaller and the arm longer and flatter. Whether this difference in construction suggests a different mode of bipedal locomotion—different stance, less energetically efficient, more efficient—remains an open question. There is, however, a growing consensus that these anatomical differences probably had only minimal behavioural consequences.

Like *Homo* species with which they are contemporary, *Australopithecus africanus* and *A. robustus* were probably never far from trees, although the more robust species was likely to be less adept in the branches than the gracile *africanus*. Although no good example of australopithecine hand bones have been recovered from this period, it is safe to conclude

The cranium of a gracile australopithecine, *Australopithecus africanus*, from Sterkfontein, South Africa (left) and of a robust australopithecine, *Australopithecus boisei*, from Lake Turkana, Kenya (right). Note the more massive face, saggital crest and pronounced zygomatic arches in *boisei*: these are all anatomical adaptations to massive musculature that powered the larger mandible in this large australopithecine species. Courtesy of Peter Kain and Richard Leakey.

from the Hadar material that their structure allowed relatively fine manipulative skills. Whether the australopithecines made any of the crude stone tools that begin to appear in the record from a little over two million years ago is difficult to answer. There is, however, circumstantial evidence from one of the South African caves that indicates that stone tools appear coincidentally with the arrival of Homo, but not before. There is, of course, an inevitable inclination to suspect that the tool-makers of the time were *Homo*, the possessors of substantially bigger brains.

15 / Homo habilis

One of the most important developments in human evolution was the dramatic expansion in brain size, which, according to the fossil data available so far, began to arise two million years ago. Certain specimens recovered from deposits close to two million years old in East Africa, specifically at Koobi Fora in Kenya and Olduvai Gorge in Tanzania, apparently have cranial capacities in excess of 650 cm^3 and close to 800 cm^3. Formally, these specimens are taken to represent the first appearance of our own genus, and are termed *Homo habilis*.

Certain thigh bones recovered from two million-year-old sediments at Koobi Fora appear to be rather more human-like, in having a larger head and shorter, rounder neck than those attributed to australopithecines. These have been said to be remains of *Homo habilis*, as has a fragment of modern-looking pelvis. And from Olduvai, there are fragments of a putative *Homo habilis* foot and hand, specimens that have recently yielded some important insights into the mode of locomotion and life style of this early ancestor.

Seen from a distance, *Homo habilis* and *Australopithecus africanus* would probably have been rather difficult to distinguish. The two hominid species were of similar height and bulk (around 1.3 m tall and weighing about 40 kg), and both walked bipedally. Close-up, however, *Homo habilis* would be seen to be distinctly less ape-like around the face and cranium. The cranium was higher and rounder, and the face flatter, less protruding and less buttressed.

While the australopithecines' dental apparatus can be described as apparently adapted to a grinding function, that of *Homo habilis* was, to some degree, more refined. The cheek teeth, or molars, while still being relatively large and covered by a thick layer of enamel, were somewhat narrower than in *africanus*. The same type of reduction was even more apparent in the premolars. The incisors, however, were larger and more spade-like, apparently forming a row of slicing teeth with the relatively flat canines. The overall impression of the dental array is one of the dietary shift compared with the australopithecines. The shift cannot have been very dramatic, however,

because microwear patterns examined so far on *Homo habilis* teeth still indicate a rather generalized frugivorous diet, like that of a modern chimpanzee and like the australopithecines. The class 'generalized frugivore' is, however, quite broad and could have included distinctly different diets, but not ones that, for example, include substantial meat-eating or foraging for ground tubers.

The species name *Homo habilis* means, literally, handy man, a name chosen in 1964 by Louis Leakey, Philip Tobias and John Napier. Mary Leakey had found a shattered skull of a robust australopithecine at Olduvai Gorge in 1959. The skull had been dated at about 1.75 million years, a time frame from which many stone tools had been recovered from the Gorge. When in the early 1960s certain cranial and postcranial bones of a larger brained, less bulky hominid were found, Louis Leakey and others

Cranium of KNMER 1470, *Homo habilis*, reconstructed from many fragments found on the eastern shore of Lake Turkana, Kenya. Note the less protruding face and rounder, larger cranium than seen in the australopithecines. 1470 has a cranial capacity of about 750 cm^3 and is dated at 1.9 million years. Courtesy of Peter Kain and Richard Leakey.

made the obvious assumption that it was this presumably more intelligent creature that made the tools. Hence the name, handy man. Specifically, Napier argued that the hand bones attributed to *Homo habilis* would have been capable of a precision grip necessary for the manipulative skills deployed in stone tool-making.

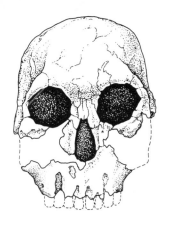
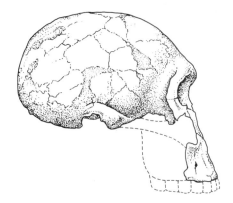
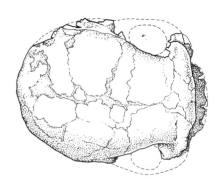

5 cm

Three views of the *Homo habilis* cranium KNMER 1470 from Koobi Fora in Kenya. Note the rounder cranium compared with the australopithecines. The postorbital constriction (the distance between the temples) is less pronounced here due to brain enlargement, about 750 cm³. From *The Hominids of East Turkana* by Walker A. and Leakey R.E.F.

A decade after the Olduvai *Homo habilis* remains were discovered there came the remarkable find of a shattered but relatively complete cranium of an apparently similar hominid from Koobi Fora. This specimen, which is generally known as skull 1470, came from deposits now dated at a little less than two million years. (Originally, the date was thought to be rather older.) The brain of 1470 measured almost 800 cm³ in volume, which puts it significantly higher than that of the australopithecines.

Hominid remains that might well be representative of the *Homo habilis* grade have been found in South Africa, specifically in the Sterkfontein cave, but they are rather fragmentary and difficult to date with any accuracy.

There appears to be no question that *Homo habilis* walked upright: the anatomy of the femur and pelvis attests to this. Recently, however, Randall Susman and Jack Stern have given this matter detailed attention, with interesting results. They studied the anatomy of the hand, foot and leg bones from Olduvai. The leg bones, the tibia and fibula, are said to articulate with the footbones, and might, therefore, come from the same individual. Moreover, they conclude that both the hand and foot bones are from a subadult and, while they were not recovered from precisely the same location in the Gorge, might have derived from the same individual. In any case, the two sets can legitimately be analysed together.

The hand bones, while displaying many characteristics of modern humans, are somewhat curved in

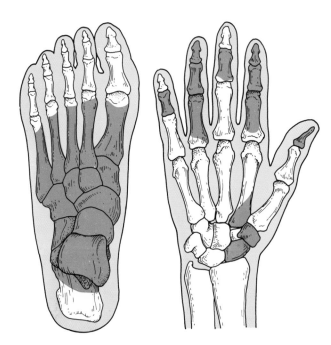

Reconstruction of the hand and foot fossil bones (shaded) of *Homo habilis* from Olduvai Gorge, Tanzania. The foot has all the appearance of modern form, though, because of their absence, nothing can be said of the toes. The hand is essentially modern too, but the curvature in the finger bones and the heavy point for muscle attachment are both evidence of a powerful grip, possibly used in frequent arboreality. The vasculature and nerve supply to the finger tips is greater in *Homo habilis* than in earlier hominids, indicating a greater sensitivity and manipulative skill.

places and more robust than in *Homo sapiens*. Stern and Susman conclude that the individual would have had a powerful grasping hand 'similar in overall configuration to chimpanzees and female gorillas'. It was a hand well suited to climbing trees.

The leg and foot bones have characteristics that are both ape-like and human-like, but overall they are much closer to those of modern humans than to apes. Stern and Susman conclude that the leg and foot 'were those of a habitual biped'.

The robust, power-grip features of the hand are not merely vestiges of an arboreal ancestor. They are the unmistakable signs of a frequent tree-climber, say Susman and Stern. *Homo habilis*, therefore, probably spent much of the day walking bipedally on the ground, with frequent excursions into the trees, perhaps in search of food, perhaps to seek safety when predators threatened. More than likely, when darkness fell, the troop of *Homo habilis* that would have foraged together during daylight hours would have retired to the safety of the trees for the night's sleep, just as modern savannah baboons do today.

The contemporary occurrence of *Homo habilis* fossils and crude flake stone tools begs the question of what the tools were used for. There is no suggestion in the archaeological record of two million to around 1.6 million years ago (the time span of *Homo habilis*) that this early form of *Homo* engaged in sytematic meat-eating. Sharp flakes might have been used for infrequent and opportunistic access to red meat on a scavenged carcass. Hide and tendons might also have been valued by those hominids. The adoption of significant exploitation of other animals as a source of food apparently did not occur until the origin of the next species in our genus, *Homo erectus*, which evolved around 1.6 million years ago.

16/The Early Hominids: an Overview

As we have already seen, the first hominid species arose some time between ten and five million years ago, the later date being favoured particularly by some of the molecular data. According to current views, by two million years ago there were at least three contemporaneous hominid species in Africa, which are allocated to two genera, *Australopithecus* and *Homo*. One million years later the hominid family had been pruned, leaving just one lineage, *Homo*, which was represented by just one species, *erectus*. Many of the more interesting questions about human origins focus on this early period of proliferation and pruning.

Judging by the relative numbers of fossils recovered from the eastern shores of Lake Turkana in Kenya, the hominids of this period were about as common as baboons are on the woodland savannah today. This confirms something about the lives of these animals that is known from other data but is often overlooked, especially in the case of *Homo habilis*. The hominids of two million years ago were not carnivores, otherwise their occupation density would have been much lower. This situation begins to change with the origin of *Homo erectus*, which, according to several sources of data, appears to have developed meat-eating as a serious dietary habit for the first time in the hominid line.

The microwear patterns on the tooth surfaces of the australopithecines and *Homo habilis* indicate that these creatures ate similar, if not the same, kinds of diet. The pattern is similar to that of the modern chimpanzee and is classified, as mentioned earlier, as principally frugivorous. In fact, the chimpanzee eats a range of food, including soft and hard fruit, succulent shoots, insects, larvae, birds eggs and, on occasions, infant monkeys and other small animals. Several different diets that include all these items, but specialized in different parts of the food spectrum, could produce the 'typical' chimpanzee microwear pattern. It is, therefore, possible that if the two australopithecines and *Homo habilis* inhabited the same locality at the same time, they would have been ecologically or behaviourally separated by their different dietary preferences. This is an important consideration because, following the principle of competitive exclusion, it is not possible for several similar species sharing very similar habits to share the same location for very long.

There is some evidence that the robust australopithecine and *Homo habilis* occupied more open, less moist environments. Indeed, both species appear to have arisen during a period of increasing aridity in Africa. Perhaps, therefore, the early hominids did not all compete directly with each other, but were geographically separated by different environmental requirements. The fact that, in some cases, fossils of different types of hominid are found in the same

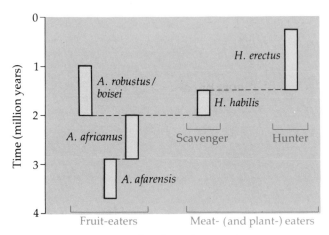

Ecological separation in early hominids. Evidence from toothwear and structure, and from archaeological material, separates the early hominids as shown. Although the australopithecines are grouped together as fruit-eaters they almost certainly occupied separate ecological niches, differentiated perhaps by overall food preferences and habitat requirements.

geological strata might simply reflect the very coarse time resolution of such sedimentary accumulation. A shift from moist to dry conditions, and back again, could occur over a period of a few hundred or even thousand years, yet might be indecipherable in most sedimentary deposits.

Nevertheless, it is reasonable to assume that the lives of three different hominid species were distinct in some way, whether this was through geographical, ecological or behavioural separation.

The appearance of stone tools in the archaeological record is roughly contemporaneous with the appearance of the larger brained hominid, *Homo habilis*. The obvious conclusion, therefore, is that *Homo habilis* made the tools, as its presumed descendants *Homo erectus* and *Homo sapiens* did. The use of stone tools almost certainly had to do with new ways of gaining food and perhaps the inclusion of new types of food. The *Homo habilis* economy was therefore likely to be more complex than that of the australopithecines, as was its social life. An increase in brain size is frequently associated with new levels of complexity in a species' way of life, as often happens when new approaches to predation evolve in carnivores. The enlarged brain of *Homo habilis* was probably associated with such an increase in economic and social complexity, perhaps involving significant sharing of food for the first time in primate history.

There is no reliable way of knowing the social structure of our hominid ancestors; however one can look for guidance from contemporary models. Most primates are intensely social and live in groups in which the mother–infant relationship forms the central bond. A period of infant dependency and learning can be presumed for the hominids, just as occurs in modern, non-human primates. And this period probably began to be substantially increased in *Homo habilis*, because the larger brain would have

meant that infants were born at an earlier stage of mental maturity. This would have had an important impact on overall social organization and habits.

Baboons and chimpanzees live in groups that may be quite small, around a dozen individuals, or very large, up to a hundred. By contrast, social groups of modern hunting and gathering people are typically around 25 individuals, primarily because of the greater territorial requirements of systematic hunting. There is evidence from the archaeological record of 1.5 million years onwards that hominid social groups tended toward the hunter–gatherers' magic number of 25, presumably for the same ecological reasons. In the era of *Homo habilis*, however, while there is occasional evidence of 'camp sites' or home bases, there is no persuasive reason to invoke a hunter–gatherer type of organization. The baboon/chimpanzee model is probably more appropriate for the australopithecines and *Homo habilis*.

One important feature of the evolution of the *Homo* lineage is the increasing development of

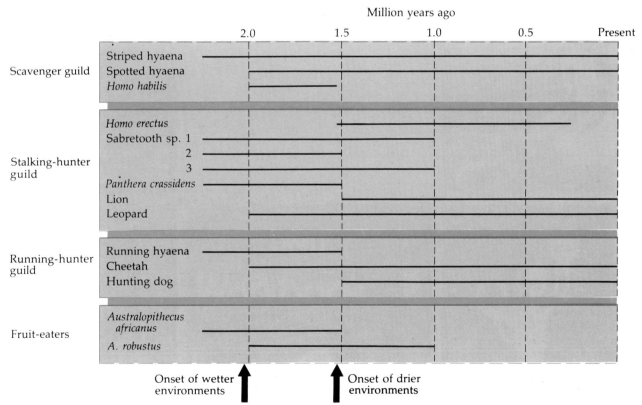

Events in hominid evolution in the context of the large carnivores. Two million years ago, East African environments became wetter, a change that saw a great increase in herbivore species and population numbers (including *Australopithecus robustus*). In response, two new species of hunter arose (cheetah and leopard) and two species of scavenger (spotted hyaena and *Homo habilis*).

Drier environments 1.6 million years ago saw the loss of a scavenger (*Homo habilis*); the loss of three species of hunters was balanced by the appearance of three new species (including *Homo erectus*). *Australopithecus africanus* became extinct at this time while *A. robustus* persisted until 1.0 million years ago, a time when the modern carnivore guild was established. Courtesy of Alan Walker.

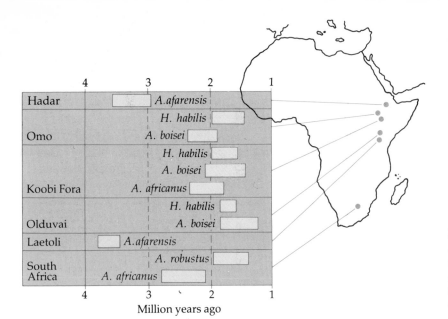

Major sites of the early hominids. Although hominids younger than one million years have been found outside Africa, so far none has been discovered with a confirmed older date than this.

opportunism and adaptability: the diet and ecological exploitation. This, together with the rapidly evolving and spreading woodland and open country baboons, was probably the cause of the australopithecines' extinction, through competitive exclusion.

It is worth noting at this point that in the study of human evolution there has always been a propensity to erect a relatively neat ladder of progression from one form to the next, slightly improved, form. However, if human history has followed the rules that operate among other animal groups, and there is no reason to suspect otherwise, then there are likely to have been more rather than less species in our ancestry. New groups, or clades, usually proliferate in number of species and genera. Instead of a ladder of human evolution, one should instead think of a bush, each branch representing a new species that, with the exception ultimately of one, became extinct. The representation of a biological family by just one species—*Homo sapiens* is the only extant hominid—is a rarity in Nature, it should be noted.

17 / Homo erectus

Homo erectus first arose at least 1.6 million years ago and continued for more than a million years before the transition to *Homo sapiens* occurred. During this long history there appeared a number of important 'firsts', including: the first evidence of systematic and cooperative hunting, as opposed to opportunistic scavenging; the first use of fire; the first systematic tool making, as opposed to opportunistic stone knapping; the first firm evidence of frequently formed and seasonal home bases, or camp sites; and (almost certainly) the first hominid occupation outside Africa. Clearly, *Homo erectus* was capable of a life more complex and varied than had previously been possible.

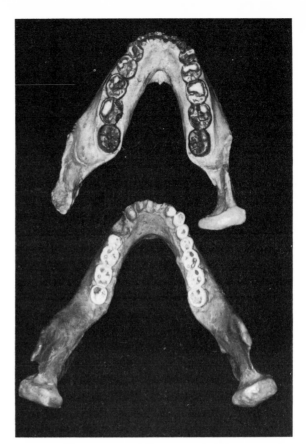

Homo erectus lower jaw (bottom) showing dentition in comparison with the robust australopithecine, *Australopithecus boisei* (top). Courtesy of Milford Wolpoff.

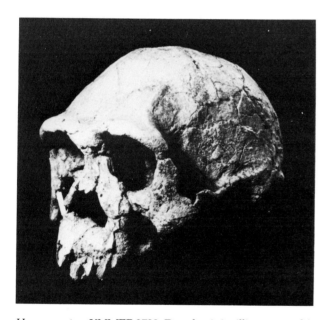

Homo erectus, KNMER 3733. Dated at 1.6 million years, this specimen is the most complete and oldest of its type so far known. Note the prominent brow ridges and rounded cranium (about 850 cm³ capacity). Courtesy of Peter Kain and Richard Leakey.

Homo erectus had a large brain, measuring between 800 and 900 cm³ in some of the early specimens to 1100 cm³ in later populations. Although still somewhat robust, the dentition was distinctly more human than in *Homo habilis*, particularly the incisors, which typically were flatter and more spade-like.

The jaw was robust and had no prominent chin, unlike modern humans. The forehead was low and sloping and, most curious of all, the brow ridges above the eyes were characteristically prominent. Given the wide geographical distribution eventually achieved by *Homo erectus*, spreading as it did throughout Africa, Asia and Europe, it is perhaps not surprising that distinctive geographical variants arose. Characteristic differences in the architecture and robusticity of the skull and the shape of anterior teeth, for example, appeared in some populations. Below the neck, *Homo erectus* was essentially human, except in the substantial robusticity of the limbs and muscle attachment points and in having a slightly shorter stature. Not surprisingly, *Homo erectus* was a habitual biped.

The very short, fine hair covering of modern humans, and the accompanying very rich distribution of sweat glands, is unique among primates. The question of when this occurred will probably never be solved, simply because the fossil record typically does not preserve soft body parts. Some

Petralona, Greece [0.3??]

Zoukoutien (Peking Man) [0.5–0.2]

Steinheim, Germany [0.25]

Swanscombe, England [0.25]

Vertesszöllös, Hungary [0.5]

Arago, France [0.25]

Lantian, China [0.75]

Sale, Morrocco [0.25]

Ternifine, Algeria [0.7]

Awash, Ethiopia [0.3]
Koobi Fora, Kenya [1.6]
Laetoli, Tanzania [0.13]

Sangiran and Trinil [0.75]

Swartkrans, South Africa [1.0??]

Modjokerto [1.5??]

Olduvai Gorge, Tanzania [1.25]

Major *Homo erectus* sites. Figures in parentheses indicate age; when followed by ?? the date given is still unsettled. Note that all the sites older than one million years are confined to Africa, with the exception of the highly uncertain attribution of 1.5 million years to Modjokerto in Indonesia. The dispersal of *Homo erectus* from Africa to the rest of the Old World might be related to the greater territory required, and allowed, by an active predatory habit in a large primate.

people have argued that the loss of thick fur was an adaptation to an energetic life on the savannah, a way of preventing overheating in a hunting primate. There are many reasons why this is unlikely, and perhaps a more plausible explanation is that the reduction in thickness of hair cover is not a direct adaptation at all but instead is the by-product of other important changes. For example, resetting of the developmental clock that led to the enlarged brain might also have retarded the growth of body hair. In which case, our ancestors might have started to become 'naked' with the origin of *Homo habilis*, a trend that would have been accentuated in *Homo erectus*. Selection for body temperature control might have been secondary.

In any case, it is safe to assume that, in common with most tropical primates, the skin colour of the early hominids was dark, and it would have become darker with the loss of the protective layer of fur.

Homo erectus remains have been discovered throughout Africa, Europe and Asia. With the notable exception of one site in Indonesia and one in China, which have unconfirmed dates of 1.9 and 1.7 million years respectively, all the early specimens have come from Africa. Sites in Eurasia are usually one million years old or less. The simplest story of human history would have hominids confined to Africa until 1.6 or so million years ago, when the more adaptable, more resourceful and, most impor-tant, more carnivorous *Homo erectus* would eventually have been capable of exploiting territories further from the tropics. This story may turn out to be too simple.

One of the richest sources of *Homo erectus* fossils is

the famous Zoukoutien Cave near Beijing, China. Fragments of at least 40 individuals of 'Peking Man' have been recovered from this site, most of which were tragically lost at the outbreak of World War II. The fossils apparently documented an increase in brain size from about 900 cm³ in specimens from the oldest deposits, dated at 600 000 years, to about 1100 cm³ in the youngest specimens, aged about 200 000 years. Also documented is the sytematic use of fire. Ash accumulation from one of the hearths eventually reached 6 m thick. Evidence for the early use of fire comes from Yunnan Province elsewhere in China, near to Marseilles in France and Vertésszöllös in Hungary. The earliest putative use of fire, however, comes from a 1.4 million year old site at Chesowanja

in Kenya, a claim over which there is still some dispute.

One of the hallmarks of *Homo erectus* was a particular type of stone tool, the tear-drop shaped hand axe. These implements, which are usually called Acheulian hand axes after a site in France where they were first discovered, are found in 1.4 million-year-old deposits at Olduvai Gorge contemporaneously with the appearance there of *Homo erectus* specimens. Older specimens have been recovered from Koobi Fora in Kenya. Hand axes are found— sometimes crudely made, sometimes beautifully fashioned, and sometimes showing indications of individual or local styles—from these early dates right through to 200 000 years ago. Curiously, how-

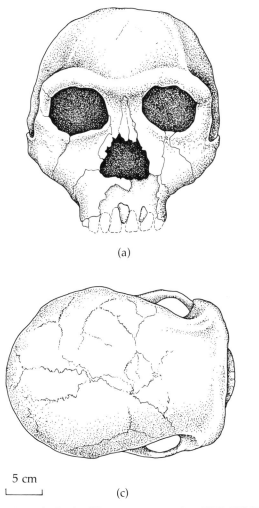

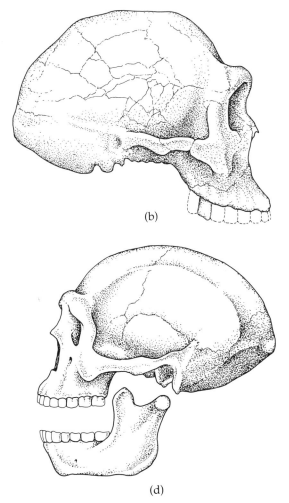

Three views (a, b, c) of *Homo erectus* cranium KNMER 3733 from Koobi Fora in Kenya. The evolutionary expansion of the brain seen in *Homo habilis* is continued in this species, being about 850 cm³ in this specimen. The pronounced brow ridges clearly distinguish the specimen as *Homo erectus*. Note that the skull comes almost to a point in the occipital (back of head) region, a feature seen in some other

populations of *Homo erectus*, particularly in Peking Man (d). The Peking population lived one million years after those at Koobi Fora. Adapted from Le Gros Clark W.E. (1955) *The Fossil Evidence of Human Evolution.* p. 195. Chicago University Press; and from *The Hominids of East Turkana* by Walker A. and Leakey R.E.F. Copyright © (1978) by Scientific American, Inc. All rights reserved.

55

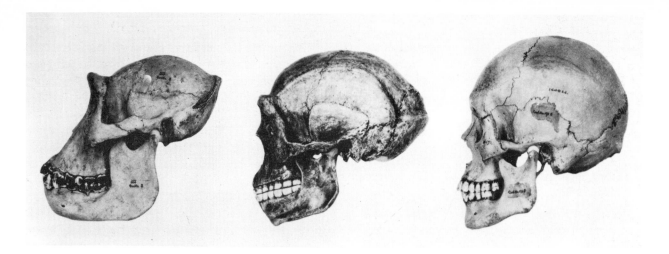

Skull of *Homo erectus* (Peking Man) from Zoukoutien, China (centre) compared with skulls of a gorilla (left) and modern Chinese male (right). Courtesy of American Museum of Natural History.

ever, they were not an important feature of tool technologies in East Asia.

There are signs of hominid 'camp sites' from very early in the record, but it is only after the origin of *Homo erectus* that they begin to appear with any significant frequency and with sure remnants of occupation, as opposed to a briefly used butchery site, for example. Just as the frequency of camp sites increases in the archaeological record, so does their sophistication. A scatter of fractured bones and stones on a briefly occupied 1.5 million-year-old river bank in Kenya is typical of 'camp sites' of the period. Compare this with a seasonally constructed 3×6 m hut built 300 000 years ago on the northern shores of the Mediterranean, in Nice, France. The trend in the record suggests the increasing importance through time of a home base that was the focus of social and economic activity.

Signs of *Homo erectus*, the serious hunter come from two sources. First, there are a number of bone accumulations that were clearly the result of repeated, organized hunting. A large collection of giant baboon remains, together with hundreds of hand axes, at a half-million-year-old site at Olorgesailie, Kenya, is one example. Second, the removal of meat from bones using stone flakes and cleavers leaves characteristic cut marks on the surface of the bone. The patterns of butchery marks becomes ever more systematic as time passes, suggesting an increasing commitment to routine meat eating.

The evident complexity of the life of *Homo erectus* must have placed great demands on individuals as intelligent, socially interacting beings. It would be surprising if a relatively complex spoken language had not, by this time, evolved. The prehistoric record is, of course, silent on this point.

18 / The Food-sharing Hypothesis

The anatomical changes associated with human evolution are rather straighforward and simply described: from an ape-like, tree-climbing ancestor, hominids became terrestrial bipeds that, in the surviving line at any rate, also acquired a substantially enlarged brain. The changes in behaviour associated with human evolution, on the other hand, are much more complicated and not so simply described. It is, of course, this shift in behaviour pattern that is fundamental to human origins, while the anatomical modifications are in a sense secondary. Unfortunately, it is much more difficult to reconstruct ancient behaviours from the meagre remains of the prehistoric record than it is to track, for example, an increase in brain size.

Glynn Isaac, an archaeologist at Harvard University, has constructed the food-sharing hypothesis as a model with which to shape his investigations of the archaeological record of the past two million years. The model, which currently enjoys wide popularity, proposes that food-sharing among early hominids was the central feature in a major restructuring of social and economic life compared with that of apes.

Isaac lists six key differences between modern humans and modern apes. First, humans are bipedal and habitually carry around tools, food and other items. Second, while apes clearly communicate with each other, none of them does so using a spoken language. Third, the acquisition of food among humans is a corporate responsibility, an activity involving exchange and sharing among adults and young; food acquisition among apes is a solitary affair. Fourth, humans typically postpone consumption of some food until they return to a home base; apes eat the food where they find it. Fifth, although chimpanzees occasionally hunt, their prey is usually small whereas human hunters can capture large animals. Sixth, primitive hunter–gatherers employ a small but effective collection of tools for their subsistence activities, such as stone flakes, digging sticks, spears and crude containers; apes do occasionally use blades of grass or twigs while 'fishing' for termites, but they have never been seen to employ tools while hunting or eating meat.

Many of these differences are clearly very significant, especially the possession of spoken language. However, as Isaac points out, the differences in the realm of meat-eating and tool-using are differences in kind, not in nature. Humans' food-sharing habit and postponement of consumption are, however, distinctive and they require the establishment of a home base, which is unlike anything in the daily lives of apes.

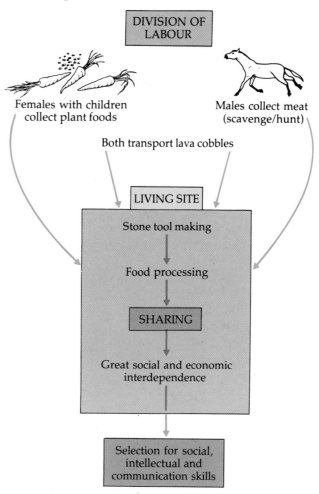

Components and relationships of the food-sharing hypothesis. Division of labour in the collection of various types of food, and their subsequent pooling and sharing at a living site, forms the core of the food-sharing hypothesis. This complex means of subsistence produced a great social and economic interdependence not previously elaborated by primates. It would have generated a sharp selection pressure for intellectual and communicative skills.

Archaeological evidence for some kind of food-collection, food-sharing focus comes as early as two million years ago, both from Olduvai Gorge and Koobi Fora. In both locations there are spatially concentrated collections of crude stone tools and bones of many different types of animals. The impression is of a site, a home base, to which whole or parts of animals were brought over a short period of time. And in several instances, the stone at the site must have been transported from some distance away as no local source was available.

The hominids were doing more than simply carrying things around. They apparently were focusing their activities on certain specific sites. The question is, Why? Why were the hominids behaving in an un-ape-like manner? Some have suggested that while adult hominids were out collecting food, the

their spoils back to camp, where they are shared out, either among the immediate family or more widely among the group. This behaviour is, of course, that of *Homo sapiens*, not necessarily of earlier hominids. But, argues Isaac, the archaeological record is consistent with the incipient development of such a system as long ago as two million years. By one million years ago there is little doubt that it was firmly underway.

One very practical benefit accruing to a group committed to division of labour and food-sharing is

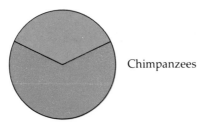

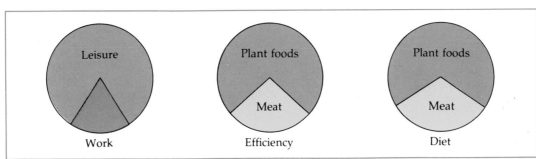

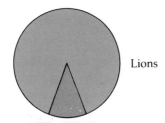

Subsistence effort in hunter-gatherers. Studies on the !Kung San (bushmen) of Botswana have shown that the combination of hunting and gathering is a highly efficient subsistence economy. The 'pie slice' on the left indicates that, on average, !Kung adults spend 2.25 h/day in the quest for food (compared with chimpanzees and lions); the middle pie slice shows that for the same amount of effort, gathering produces 2.5 times as much food as does hunting; the figure on the right shows the composition of the diet. Once hunting and gathering became established, perhaps beginning as long as 1.5 million years ago, it remained the primary human adaptation, with all its social consequences, until the advent of agriculture, 10 000 years ago.

young were left in safety at a home base, an explanation for which there are precedents in other animals. Isaac argues, however, that the home base is the product of a division of labour, which is inextricably linked with the habit of food-sharing.

Among recent hunter–gatherers, males typically hunt while females, who are encumbered with young, typically gather plant foods. Both sexes bring

strictly economic. By diversifying subsistence activities in this way, the range of foods that can be collected is expanded substantially. As subsistence patterns are fundamental to an animal's success, a new pattern that produces greater economic security could well have been an engine for human evolution.

As several adventurous experimenters have demonstrated, the savannah offers a generous

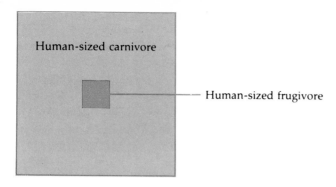

Human-sized carnivore

Human-sized frugivore

Relative subsistence areas. A human-sized carnivore might require 20 km² or more from which to derive daily subsistence compared with perhaps 0.5 km² for an equivalent-sized fruit-eater. A shift to active predation in *Homo erectus* would, therefore, dramatically affect the population structure and dynamics of hominids.

supply of disabled animals and abandoned young to the vigilant scavenger. The scavenger, must, however, be able to gain access to the meat, which is generally protected by tough hide. The invention of the simple, but highly effective, stone flake would therefore have been a major technological innovation that would have given access to a rich supply of food previously untapped by any primate. The development of hunting skills would have further expanded this food source.

As has been stated, the archaeological record of two million years ago attests to some form of activity focus involving animal bones and the making and use of stone tools, although the pattern becomes common and clearly defined only later, during the era of *Homo erectus*. The record is silent, however, on the use of plant foods, for the very good reason that such material is virtually certain never to enter fossil record. If patterns of subsistence among recent hunter–gatherers is any guide-line, these early hominids would have eaten a diet principally of plant foods supplemented by occasional bonanzas of meat, either hunted or scavenged. The gathering of plant foods and their transport back to a home base would have required simple technology, such as a crude container, perhaps made of bark or animal skin. Sharp digging sticks would give access to nourishing and succulent roots and tubers, an innovation that would not only have widened further the dietary possibilities for early hominids but also would have permitted subsistence in areas devoid of food on bushes and trees.

The economic benefits of division of labour and food-sharing would clearly have revolutionized the social structure of the early hominids, making it much more complex. For example, the exchange and sharing of food, Isaac points out, would have sharpened the sense of reciprocal obligation between individuals, both kin and non-kin. The social and economic unit becomes tightly knit and would eventually have come to function with a complexity unimaginable in the absence of an equally complex spoken language.

19/Campsite Reconstruction: the Test of a Hypothesis

Archaeological excavations typically are thorough, painstaking projects in which the exact coordinates of every recovered item, bone or stone, are meticulously recorded. Because of the many agencies that can influence the distribution and accumulation of bones and stones (see section 10), it is often not immediately clear that an excavation in very ancient deposits is uncovering the remains of a campsite or merely the remnants of a randomly formed assemblage.

It is, therefore, incumbent on those who excavate putative living sites to look for patterns in the material recovered that indicate without reasonable doubt the work of protohuman hands. It is also important that archaeologists should attempt to answer, through their excavations, specific questions about protohuman behaviour. One very good example of a project conforming with these strictures was an excavation of FxJj50, otherwise known as Site 50, at Koobi Fora in Kenya, a project led by Harvard archaeologist Glynn Isaac. So convincing was the evidence of hominid activity at the site that when Isaac and his colleagues (Henry Bunn, John W. K. Harris, Zefe Kaufula, Ellen Kroll, Kathy Schick, Nicholas Toth and Anna K. Behrensmeyer) reported their work in the scientific literature they wrote: 'During the excavation of Site 50 we have had a privileged sense of gaining glimpses of particular moments in the lives of the very early protohumans who lived in East Africa 1.5 million years ago.'

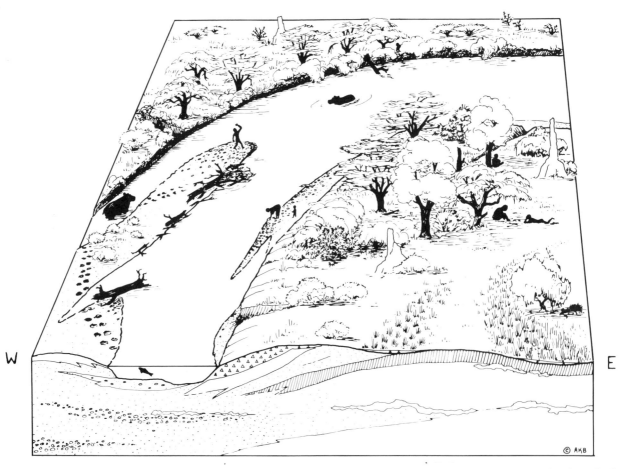

Sketch by A.K. Behrensmeyer indicating the palaeotopography of Site 50 and the stratification that provides evidence for the reconstruction. The hatched layer is the tuffaceous sandy silt in the top of which the archaeological horizon is stratified. Examples of some tree genera represented in the pollen are shown and marked: C = *Commiphora*, F = *Ficus* and S = *Salvadora*. Other trees are three species of *Acacia*. Courtesy of A.K. Behrensmeyer.

A small core/unifacial chopper with hammerstone in position to knock off next flake (from Site 50). Courtesy of Glynn Isaac.

The excavation of Site 50 was undertaken very deliberately to test some aspects of Isaac's food-sharing hypothesis (section 18). Specifically, what evidence is there that hominids 1.5 million years ago transported bones and stones to a favoured location where both would be processed, the stone to provide tools and the bones food?

When hominids were living in the area, Site 50 was located in the middle of a large floodplain on the eastern shore of Lake Turkana. The landscape was typical open savannah, with short *Acacia* and *Commiphera* trees scattered over the terrain, while thicker groves of taller trees lined the water courses that laced the floodplain. Plains animals, such as giraffe, zebra, antelope and baboons, lived there. Site 50 was formed on a sandy bank in the crook of a winding river, a location that would have offered access to water, shade from the sun, a supply of fruit and berries from nearby bushes, and ready access to lava cobbles suitable for tool-making.

During the two-year long excavation, 1405 stone fragments and 2100 pieces of bone were recovered, distributed in a thin layer over an area of about 200 m². The density of bones and stones within the site was more than ten times higher than outside the putative campsite area. Bones and stones within this area were concentrated into two distinct spots, suggesting two locations of particular activity.

Less than one-half of the bone fragments could be positively identified, but it was clear from those that could be that a wide range of animals was represented at the site, albeit by just a few bones in most cases. Remains of every major group apart from carnivores, rhinoceroses and elephants were present. One particularly interesting specimen was the shattered shaft of a leg bone (humerus) of an eland-sized antelope, *Megalotragus*. By careful reconstruction, Henry Bunn was able to fit seven large fragments together, which revealed fracture damage of the sort inflicted by hunter–gatherers when in search of marrow from long bones. Moreover, Bunn found that the end of a leg bone from the same type of animal (and possibly the same individual) bore a set of short narrow marks such as would be made by a sharp instrument used to deflesh the bone. Experiments with sharp stone flakes and modern bones, together with microscopy of the fossil bones, confirmed the presence of 'cut marks' on at least half a dozen major bones at Site 50.

The very small proportion of bones bearing signs of hominid activity is not at all surprising, especially in view of the likely fate of the site once abandoned.

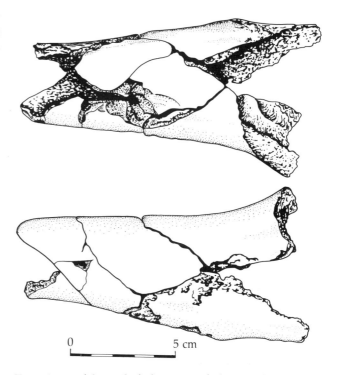

0   5 cm

Two views of the end of a humerus of a large extinct antelope that had been shattered, presumably by occupants of Site 50. In addition to indications of percussion, this bone also bears cut marks that were probably made by sharp stone implements, such as flakes, that might have been used to remove meat or other tissues. Courtesy of A.B. Isaac.

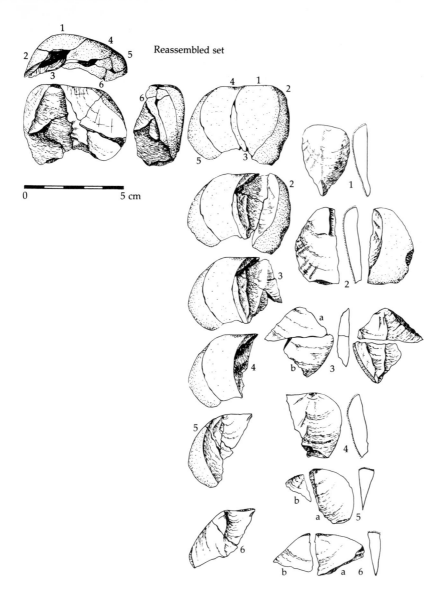

Reassembled set

0   5 cm

An archaeological jigsaw. This set of six flakes was scattered over a small area in the south east corner of the living site. They had been struck from a water-worn cobble to produce a side chopper, which was apparently removed from the site. The diagram shows two views of the six flakes (top left) as they would have been as part of the intact cobble. The order in which the flakes were stuck is seen in the descending pictures, each flake being numbered 1–6. Courtesy of A.B. Isaac.

Although geological evidence shows the site to have been quickly and gently covered by flood-transported silt, there are also unmistakable signs of scavenger activity among the remains. These include fossilized carnivore dung and gnaw marks on some of the bones. In addition to crunching some of the bones on the site, the carnivores might well have removed some or even been responsible for the presence there of still others.

Among the putative stone artifacts were 59 items that, according to traditional classification, would be labelled choppers, polyhedrons, discoids, core scrapers, scrapers or flakes and flaked core/cobble fragments. In addition, there were more than 1300 flakes and flake fragments. Although the numbers sound large, a practised stone knapper could have produced this type of assemblage in about an hour,

according to Nicholas Toth.

Another member of the research team, Ellen Kroll, spent many patient hours trying to fit fragments of stone back together, a project rewarded by the assembly of 53 sets of two or more pieces. The conjoining of a series of flakes, sometimes together with the core from which they were struck, sometimes without, gave an important coherence to an otherwise apparently jumbled heap of broken stones. The distribution of the conjoining sets and their concentration into two distinct activity areas is strong testimony to deliberate hominid activity.

The natural presumption is that the stones were gathered nearby, taken to the site and then struck in order to produce sharp flakes with which to process the pieces of carcass that were also brought to the river bank site. In a few instances, this presumption has been confirmed about as strongly as it could be with present techniques. Lawrence Keeley, at the University of Illinois, and Toth have examined the sharp edges of some of the flakes and they detect on two of them the unmistakable evidence of meat-slicing. Both of these flakes had been found within a metre of the cut-marked leg bone. Moreover, another flake from the site shows evidence of use on soft plant tissue, which is a rare signal of the presumed common use of vegetable foods by early hominids.

The size and material accumulation of the site indicated it was used by a small group of individuals for a relatively short time, perhaps just a few days. It is a reasonable conclusion from the data accumulated in the excavation and their subsequent analyses that the hominids of the time were transporting stones, parts of carcasses and plant foods to the site, where they were then processed. The range of bones on the site indicated scavenging rather than hunting as the primary source of meat for these hominids. All this is fully consistent with the food-sharing hypothesis, although other possible explanations still need to be considered and tested.

20/Tool Technology: the Early Cultures

The use of tools had a tremendous impact on the path of human prehistory, specifically in allowing access to food not otherwise available, and their beneficial economic rewards played an important part in the ultimate biological success of the ancient tool-makers. Because stone does not perish through archaeological time, whereas hide, tendons, wood and bark rarely become fossilized, the prehistoric record is heavily biased toward stone tool technology, in the early stages at least. Digging sticks and wooden spears may well have played an important, perhaps even dominant, role in early technologies, but the record is virtually silent on the matter.

ment, one will never know. But the discovery is important, not least because it shows the use of one tool in the preparation of another.

The earliest putative stone artifacts discovered so far come from Ethiopia and are dated at around 2.5 million years. They are a collection of extremely crude scrapers, choppers and flakes, each the product of a very few blows with a hammerstone. In archaeological terms, they are described as an example of the Oldowan industry. Looking forward through time from this earliest example of tool-making, one gains two powerful impressions.

First, there is a striking continuity through vast tracks of time. Tools such as these earliest artifacts represent the dominant form of stone tool technology for more than a million years. About 1.5 million years ago a new industry emerges, which is known as the Acheulian. This industry represents only a modest advance over the Oldowan, and is characterized by the presence of tear-drop shaped handaxes. The Acheulian did not replace the

Representative examples of Oldowan tools. Top row: hammerstone; unifacial chopper; bifacial chopper; polyhedron; core scraper; bifacial discoid. Bottom row: flake scraper; six flakes. An actual tool kit on site would mainly comprise flakes. Courtesy of Nicholas Toth.

Representative examples of Acheulian tools. Top row: ovate handaxe; pointed handaxe; cleaver; pick. Bottom row: spheroid (quartz); flake scraper; biface trimming flake; biface trimming flake. (All artifacts, except spheroid, are lava replicas made by Nicholas Toth). Courtesy of Nicholas Toth.

Recently, however, the work by Lawrence Keeley and Nicholas Toth on wear on stone tools from Kenya provided a useful reminder of the probably common use of wooden implements in early times. According to their analysis, one of a series of small stone flakes from a 1.5 million-year-old site shows distinct signs of woodworking, as if it had been used in whittling a stick. Whether the finished product was a spear, a digging stick or some other imple-

Oldowan immediately, but merely accompanied it through half a million years of human history, after which it became the dominant form. Even so, tools that can be described as Oldowan in type were still to be found in Eastern Asia right up to 200 000 years ago and less. In Africa and Europe the Acheulian continued as the main tool industry, until it too began to be replaced around 150 000 years ago.

The second powerful impression of stone tool

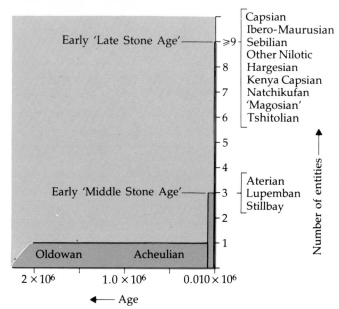

Increase in styles of tool technologies. Compared with later industries in Africa—the Middle Stone Age and Later Stone Age—the long eras of the Oldowan and Acheulian technologies each formed a single entity; there was no clear diversification into different styles. Courtesy of Glynn Isaac.

technologies up to about 150 000 years ago is their essentially opportunistic nature. Although there is a gradual imposition of form and style through that great swath of time, it is really rather minimal. Only after 150 000 years ago is there a strong sense of stylistic order.

The characteristics of the Oldowan technology have been meticulously studied by Mary Leakey through her many decades of excavation at Olduvai Gorge, the site after which the industry is named. It is a collection of perhaps half a dozen main forms in which the so-called pebble chopper predominates. Discoids, spheroids, polyhedrons, core scrapers, flake scrapers and hammerstones are some of the items in the industry, together of course with a large representation of debitage, which includes small sharp flakes. Mary Leakey is careful to point out that debitage does not necessarily mean useless waste, as the small flakes very probably were deliberately struck as stone 'knives'. The principal raw material was lava cobbles, although chart and other similar rocks were sometimes used.

The oldest levels at Olduvai date back to almost two million years ago, and it is here that the Oldowan industry begins. The industry continues for one million years and more, but becomes a little more refined, adding a few more tool categories, such as

awls and protobifaces. These advances, which appear about 1.5 million years ago, are recognized as the Developed Oldowan. At about the same time, a new industry, the Acheulian, appears in the record. The handaxe, as has been noted, is the hallmark of this new industry, and it represents the first tool in which a predetermined shape has been imposed on a piece of raw material.

The principle invention of early stone tool technologies was that of concoidal fracture: strike a core at an angle and a flake, large or small, is removed. The resultant tool is very much determined by the shape of the starting material. However, the bifacial symmetry of the handaxe, with its two sharp converging edges, required the shape to be 'seen' within the lump of stone, which is then worked towards with a series of careful striking actions.

Some handaxes of later times were aesthetically pleasing products of hours of skilled labour. Exactly what they were used for is still something of a mystery, but the combination of a long sharp edge with bulk and weight makes them exceedingly efficient at slicing through even the toughest hide, including that of elephants and rhinoceroses.

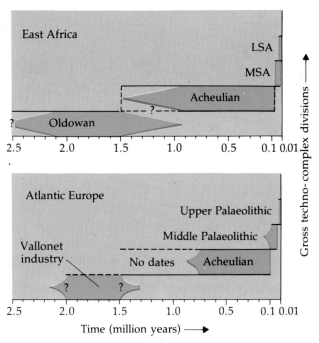

Time duration of tool technologies. With the passage of time each successive tool industry is of ever shorter duration. The element of stability in the early industries—first the Oldowan then the Acheulian—is striking. Later industries—the Middle Stone Age and Later Stone Age (MSA and LSA) in Africa and the Middle and Upper Palaeolithic in Europe—pass fleetingly by comparison. Courtesy of Glynn Isaac.

The full set of artifacts from a small site on the eastern shore of Lake Turkana. The collection shows that the majority of pieces, as in most sites, are small, sharp flakes. Courtesy of Glynn Isaac.

A classic Oldowan chopper. Courtesy of Glynn Isaac.

Cleavers, which were also part of the Acheulian technology, had similar properties, but are cruder, less elegant implements.

The Acheulian industry, which had about ten principle implements, continued for more than a million years, before it was replaced by the wide range of much more refined tools of the Mousterian culture: the products of *Homo sapiens neanderthalensis*. During that long period of time the best examples of the Acheulian technology became increasingly more elegant, but throughout this time there were crude examples, similar to those from the beginning of the record.

Glynn Isaac points out that from the very beginning of stone tool-making the range of implements produced does not increase significantly. What does change through time, however, is the degree of standardization, the frequency of producing certain forms against a background of 'noise'. *Ad hoc* stone knapping gives way to deliberate imposition of preconceived order.

The duration of the Acheulian saw certain idiosyncratic expression, for example in details of the shape of tools and their size. Differences in availability of suitable stone, different specific technological needs and an element of individual style would have contributed to this. There was, however, a certain homogeneity at any particular time. There was, in a sense, just one Acheulian culture. From 150 000 years onwards, this pattern of culture began to change, at an ever accelerating pace. It is, as Isaac says, as if some threshold was passed: 'a critical threshold in information capacity and precision of expression.'

21/Tool Technology: the Pace Changes

Between 250 000 and 150 000 years ago the pace of change of tool technologies began to accelerate. Whereas continuity was the hallmark of tool-making prior to this turning point, change began to dominate thereafter. Moreover, each succeeding culture contained a larger array of finer implements than the last. Bone, antler and ivory became increasingly important raw materials for tool-making, particularly for fine, flexible and sharp implements. And, most striking of all, there began to emerge a previously unseen degree of variability in the form of tool-kits found in neighbouring sites, a variability that has been explained variously as discrete functional differentiation or cultural expression through style.

Solutrean laurel leaf blade, some examples of which are so thin as to be translucent. They were probably used in rituals rather than in practical affairs. (Scale bar: 5 cm.) Courtesy of Roger Lewin and Bruce Bradley.

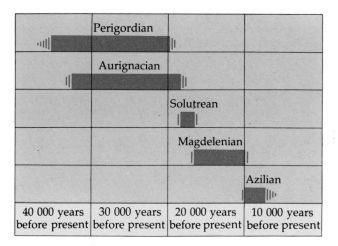

Perigordian			
Aurignacian			
	Solutrean		
	Magdelenian		
		Azilian	
40 000 years before present	30 000 years before present	20 000 years before present	10 000 years before present

Tool industries of the Upper Palaeolithic.

The major post-Acheulian tool culture was the Mousterian, associated with the origin of archaic *Homo sapiens:* this was known in Europe as the Middle Palaeolithic. The Mousterian continued through to around 35 000 to 40 000 years ago, which coincided with the emergence of fully modern humans, *Homo sapiens sapiens.* The 100 000 year tenure of the Mousterian was overtaken by a succession of tool-kits, known as the Upper Palaeolithic, that displayed the ever increasing virtuosity of the tool-makers, so much so that some implements lost any function they might have had and instead assumed some kind of abstract symbolism. (In Africa the equivalent periods are called the Middle Stone Age and the Later Stone Age.)

Towards the end of the Acheulian era, some 200 000 years ago, there arose in South Africa a new technique for the production of large flakes that was to foreshadow later developments in tool technology. Known as the Levallois technique, this new development involved a much more intensive preparation of a core than had hitherto been the practice. Virtually complete flakes could then be struck from the core at a single blow, although they were typically retouched to give the final desired shape. It was principally a refinement and development of the Levallois technique that formed the basis of Mousterian tool technology of the Middle Palaeolithic.

One immediate practical consequence of careful core preparation is a greater efficiency in the use of raw materials. For example, the basic Acheulian method yielded just 5.1–20.3 cm of cutting edge from 0.45 kg of flint, whereas a Mousterian tool-maker could strike 10.2 m of cutting edge from the same amount of starting material. This trajectory of greater efficiency soared following the origin of modern humans, 40 000 years ago, who could manufacture 12 m of cutting edge from 0.45 kg of flint, struck in the form of long sharp blades.

In addition to a greater efficiency in the use of raw materials, the Mousterians also made a much wider range of stone implements by patiently and sensitively retouching the basic flakes. The great French archaeologist Francois Bordes counted as many as 60 systematically produced categories across the whole range of the Mousterian, although at no single site would all 60 types be found. The Mousterian tool-kit included small hand axes, flake blades, scrapers and many items for detailed work, such as points, awls and burins.

Neanderthal people, *Homo sapiens neanderthalensis*, who have been particularly associated with the Mousterian culture, occupied the Old World from Western Europe through to Central Asia. Because of certain characteristic patterns associated with some of the Mousterian tool-kits, Bordes suggested that there were several different tribes of Neanderthals, each expressing their social identity through stylistic conformity. Another interpretation of the very real differences between some of the sites, advanced particularly by American archaeologists Lewis and

Middle Palaeolithic artifacts. These are typically retouched flakes of various types, made between 200 000 and 40 000 years ago. Top row, left to right: Mousterian point; Levallois point; Levallois flake (tortoise); Levallois core; disc core. Bottom row, left to right: Mousterian handaxe; single convex side scraper; Quina scraper; limace; denticulate. (Scale bar: 5 cm.) Courtesy of Roger Lewin and Bruce Bradley.

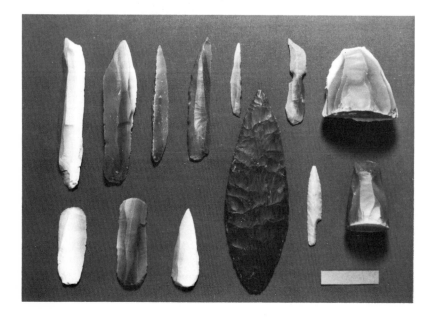

Upper Palaeolithic artifacts. These are typically formed from retouched blades and are therefore finer than Middle Palaeolithic tools. Top row, left to right: burin on a truncated blade; dihedral burin; gravette point; backed knife; backed bladelet; strangulated blade; blade core. Bottom row, left to right: end scraper; double end scraper; and scraper/dihedral burin; Solutrean laurel leaf blade; Solutrean shouldered point; prismatic blade core. (Scale bar: 5 cm.) Courtesy of Roger Lewin and Bruce Bradley.

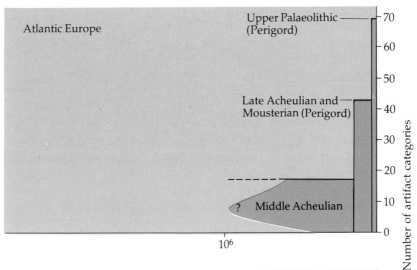

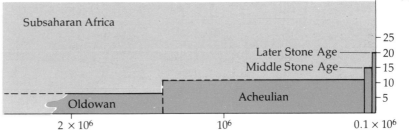

Increase in pace of new tool technologies. The emergence of new, distinct tool technologies at an ever-accelerating pace through time was accompanied by a great increase in the number of distinct implements comprising the cultures.

Sally Binford, is that they represent different sets of tools required for different functions.

It would be surprising indeed if, for example, the implements employed for woodworking were not different from those used at a butchery site, so functional differentiation must play a part in the variation seen in Mousterian archaeological sites. However, it would be equally surprising if the Mousterian people, as intelligent and sensitive as they evidentally were, did not perceive complex social structures that were differentiated through identifiable styles, in tool-making and no doubt in other ways too.

In many ways, the shift 40 000 years ago from the Middle Palaeolithic to the Upper Palaeolithic was not as dramatic as that between the Acheulians and the Mousterians. Fine narrow blades were a prominent feature of the Upper Paleolithic tool-kits, as were delicate implements of bone and antler. Tool preparation involved a greater emphasis on precisely controlled pressure flaking as opposed to a free stroke with a hammerstone. These items had been part of the Middle Palaeolithic, but they were not especially emphasized. Archaeologists recognize new tools in the technologies of these newly evolved modern humans, so that overall there were as many as 100 identifiable implements in the Upper Palaeolithic.

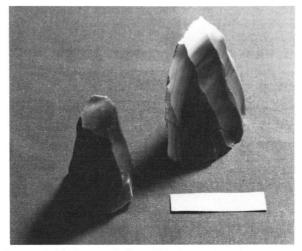

Prismatic blade core (left) and blade core (right), two sources of narrow flint blades in Upper Palaeolithic tool technology. Repeated blows around the edge of the flat surface (upon which the cores are resting in the picture) yields large numbers of sharp blades which can be finely retouched. (Scale bar: 5 cm.) Courtesy of Roger Lewin and Bruce Bradley.

These included hafted implements, which involved the conceptual and technological advance of combining two artifacts. The important point, however, is that new items arose every few thousand years

69

throughout the Upper Palaeolithic as compared with every five or ten thousand in the earlier era.

The tremendous emphasis of research on the Upper Palaeolithic in Europe has yielded a wealth of detail as compared with the rest of the world. As a result certain 'cultures' have been named as representing specific periods within the Upper Palaeolithic. For example, two principle cultures, the Aurignacian and the Perigordian, are described as having coexisted for much of the earliest part of the period, from around 35 000 years to around 20 000 years ago. As one would anticipate, there is considerable consistent variation within these cultures too, which has again engendered discussions on cultural style and utilitarian need.

The successors to the Perigordians were the Solutreans, who, among other things, were distinguished by their fine laurel leaf blades and their invention of the eyed sewing needle. The Solutreans, 20 000 to 16 000 years ago, were followed by the Magdalenians, who were responsible for the high art of Ice Age Europe.

With the passing of the Ice Age 10 000 years ago the Magdalenian culture passed too, to be replaced by an ever greater burgeoning of traditions associated with the new activities of the Agricultural Revolution. A new archaeology began to arise, the archaeology of towns and cities, of trade and of conflict, patterns that in a remarkably short period of time were to be found throughout the world, both Old and New.

22 / The Neanderthals

To call someone Neanderthal is, in current parlance, to imply that they are a slouching, dim-witted brute biologically located somewhere between an ape and a human. This unflattering set of attributes derives from an erroneous reconstruction of an arthritic Neanderthal skeleton in the first decade of this century. French palaeoanthropologist Marcellin Boule, who was responsible for the error, believed Neanderthals to be distinctly primitive and inferior to modern humans, and therefore unconsciously imposed his preconception on the reconstruction. The rest of the palaeoanthropological community were of much the same mind, and readily accepted a projection they wanted to believe. The mistake was realized in the 1950s, and since then the scientific community, if not the public at large, has come to recognize the Neanderthals as intelligent, highly skilled and resourceful people whose posture was the same as ours.

There are, of course, striking anatomical distinctions between Neanderthals and modern humans,

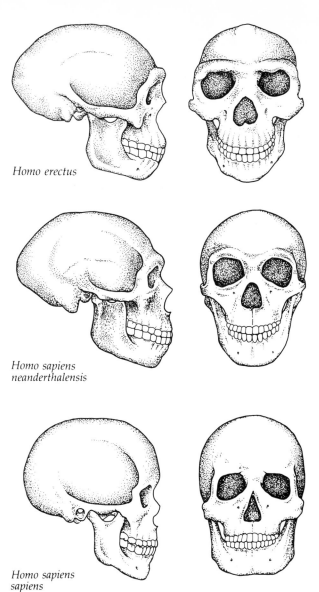

Homo erectus

*Homo sapiens
neanderthalensis*

*Homo sapiens
sapiens*

Comparison of *Homo sapiens neanderthalensis* with *Homo erectus* and *Homo sapiens sapiens* shows it to have some features of both. Neanderthal's robusticity, which is also apparent in the postcranial skeleton, echoes the physical appearance of *Homo erectus*, its presumed ancestor. The very large brain of the Neanderthals, which slightly exceeds that of *Homo sapiens sapiens*, is seen as a modern feature. Peculiar to the Neanderthals is the extreme forward projection of the face. Courtesy of Luba Gudtz.

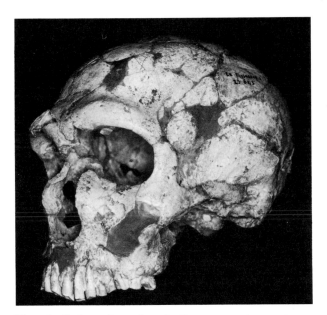

Neanderthal specimen from La Ferrassie in the Dordogne, France. Cranial capacity, 1681 cm^3. Courtesy of Milford Wolpoff.

which is why this extinct hominid form is designated as *Homo sapiens neanderthalensis* and not *Homo sapiens sapiens*. The classic Neanderthals, and some argue the only true Neanderthals, existed throughout Western Europe and across into the Near East and Central Asia from about 100000 years ago to 40000 or 35000 years ago, depending on the precise locality. No other form of human remains has been found in this area at this time.

Although the posture, range of movements and manipulation skills were the same in Neanderthals as in modern humans, the skeleton was substantially more robust. The size and cross-sectional area of limb bones, and the exaggeratedly large points for tendon attachment, indicate that these people were extremely strong, both males and females. It is clear that the muscles that go from the trunk or the shoulder blade to the upper arm were massively developed, particularly those involved in powerful downward movements. The hands, too, were powered by a very strong musculature, so much so that many of the muscle attachment points are raised as crests. In modern humans it is barely possible to detect these points of attachment.

Neanderthal legs are equally well-developed, with the thigh and lower leg bones being much greater in cross-sectional area than in modern humans. The knee and ankle joints are robust, a clear adaptation to the great stresses imposed on them by a bulky body that was so powerfully driven by a substantial musculature. These robust features are, incidentally, present in Neanderthal children and are therefore clearly an expression of these people's genetic make-up and are not extreme developments through extreme use.

On top of the Neanderthal's bulky frame was a skull that displayed a distinctive set of characteristics that set this hominid apart from others. The face is unique, in that it has a midline prominence that thrusts the nose and teeth further forward than in any other hominid, either later or earlier. The cheek arches slope backward rather than being angled as in 'high cheek-boned' modern humans. And the forehead slopes backward behind prominent brow ridges. The low cranium, prominent brow ridges and

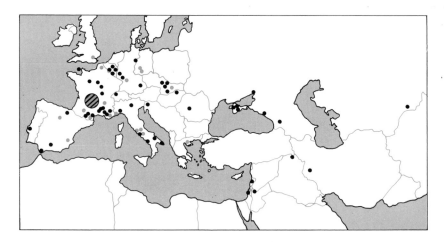

Geographical distribution of Neanderthal sites. The red dots show early sites, the black dots late sites, and the striped area a high concentration of sites, one-third of which are early and the rest late. Some robust *Homo* populations lived in Africa at this time, but most authorities limit the Neanderthal label to populations in Europe, the Near East and Central Asia.

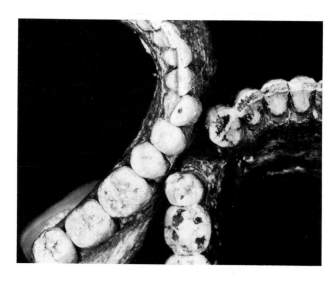

Neanderthal (right) and modern dentition showing larger Neanderthal front teeth. Courtesy of Milford Wolpoff.

robust skeleton are features reminiscent of *Homo erectus*, a similarity that has engendered the notion that Neanderthals were intermediate between the earlier hominid and modern humans. (This is an issue of some dispute, and will be explored in the next section.)

Contrary to popular conception, Neanderthals often possessed a distinct bony chin, which of course is meant to be the hallmark of modern humans. In Neanderthals, the chin's prominence was much obscured by the projecting face. The cheek teeth were about the same size as in the early *Homo sapiens sapiens*, but the incisors were larger, which tended to make the dental arch broader than in modern humans.

Neanderthal's brain was, on average, slightly larger than normal for modern humans, measuring 1400 cm³ as against 1360 cm³. Even in the absence of

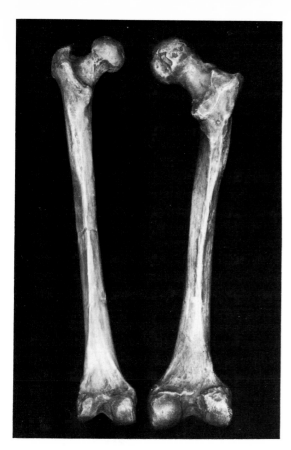

Comparison of Neanderthal (right) and modern femur. Note the robust structure of the Neanderthal femur. Courtesy of Milford Wolpoff.

however, is that the characteristic facial features were already developed before the onset of the last glacial, 75 000 years ago. Moreover, Neanderthals in the temperate Near East had the same features. So far, no special function has been firmly assigned to these people's protruding visage.

The Neanderthal people thrived under a wide range of environmental conditions, including the most harsh, almost Arctic winters of Central Asia. Where there was no wood, these people built their shelters of bones and skins, and even burned bones as fuel. They were proficient hunters, skilled toolmakers, and they made clothes.

For the first time in human history ritual burial became common. For example, at Le Moustier in France a teenage boy was layed to rest as if sleeping on his right side, his head cradled on his forearm. A pile of flints formed a pillow and a beautifully crafted stone axe rested near his hand. Bones of wild cattle were included in the burial, which suggests that these meat-laden joints might have been meant as sustenance for the boy's journey to another world.

Another famous burial was found in the Shanidar cave in the Zagros Mountains of Iraq. There, 60 000 years ago, a man was buried in the Spring, his body rested on a bed of woody horsetail festooned with a variety of flowers. This scene was reconstructed following careful pollen analysis of the soil surrounding the fossil bones. Many other examples of sensitive, ritual burial have been found, and it is clear that Neanderthal people had an acute sense of life and death, a degree of self-awareness not previously expressed in the prehistoric record.

The very distinctive Neanderthal populations lasted from about 100 000 years ago to at least 35 000 years ago in some places. The rise of the Neanderthals can, however, be detected in some European remains at least 150 000 years ago and possibly earlier. Once established, the characteristic Neanderthal form remained virtually unchanged until it abruptly disappeared 40 000 years ago in the Near East and 35 000 years ago in Western Europe.

an intact cranium for evidence, the archaeological remains of Neanderthals reveal the activities of a very intelligent, spiritually sensitive, resourceful creature. The 'extra' brain capacity was very probably required for control of the extra musculature.

The origin of Neanderthal's special features, particularly the protruding face, has long been a topic of speculation. One popular idea, which is almost certainly wrong, was that the facial projection was an adaptation to cold conditions. The nasal cavity is placed further away from the cold-sensitive brain, it was argued. The major problem with this idea,

Comparison of the Neanderthal cranium with a modern human cranium. The triangle in the Neanderthal cranium (left) shows the spatial relationships between the forward edge of the first molar (C), the lower edge of the cheekbone (A), and the upper edge of the cheekbone. A similar relationship drawn in a modern human cranium (right, with a Neanderthal outline shaded in) produces a much more flat triangle, thus illustrating the significant forward protrusion in the Neanderthal face.

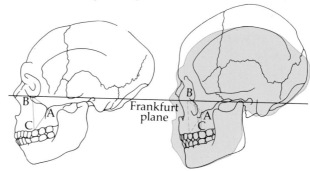

23 / Origin of
Modern Humans

The origin of fully modern humans, denoted by the subspecies name *Homo sapiens sapiens,* remains one of the great puzzles of palaeoanthropology. Current ideas can be divided at their extremes into one of two views: the replacement model and the local continuity model. The first idea envisages the evolution of modern humans from a single isolated population of archaic sapiens, which then spread throughout the Old World, replacing indigenous populations through competition or through confrontation. The second idea, local continuity, assumes that the populations of archaic sapiens that populated much of the Old World by at least 100000 years ago all independently evolved to fully sapient grade.

In this latter model the Neanderthals are viewed as simply the European and Near East representatives of relatively advanced sapiens that eventually evolved into European *Homo sapiens sapiens.* They were, in other words, a transitional form. The replacement model, by contrast, relegates Neanderthals to the status of one of many geographical variants that eventually were overshadowed by the newly evolved *Homo sapiens sapiens* from elsewhere.

There are problems with both models. For ex-ample, as University of New Mexico palaeo-anthropologist Erik Trinkaus points out, the local continuity model does not account for the genetic and morphological homogeneity of modern humans, despite their wide geographical distribution. Nor does it explain why there was an apparently rapid evolutionary transition in Europe and the Near East around 35 000 to 40 000 years ago, whereas rates of change both before and after were much slower. The replacement model, for its part, fails to explain why the earliest-known *Homo sapiens sapiens* fossils are similar to the modern inhabitants of the same regions. Moreover, these same geographical patterns of morphological variation are to be seen in archaic sapient populations.

'It has become increasingly apparent that neither model is an accurate reflection of the origins of modern humans', concludes Trinkaus. 'It appears most likely that the transition from archaic to modern-appearing humans was an extremely complicated one, involving all the known processes that come into play during periods of rapid, within-species evolution.' Perhaps some combination of the two models might be tenable: an isolated origin of the modern form, migrating populations of which inter-bred substantially with local groups they encountered. Such a scenario might account for both the high degree of homogeneity of human populations and the continuity of geographical variations.

Homo erectus, assumed to be ancestral to later sapient populations, displayed a remarkable evolutionary stability across a million or more years.

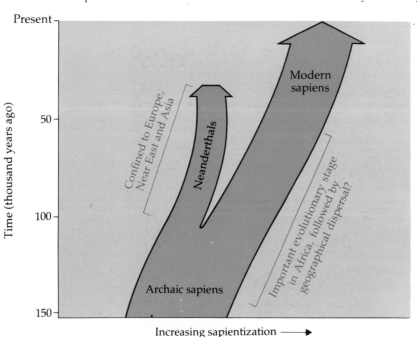

Present

Time (thousand years ago)

50

100

150

Confined to Europe, Near East and Asia

Neanderthals

Modern sapiens

Important evolutionary stage in Africa, followed by geographical dispersal?

Archaic sapiens

Increasing sapientization ⟶

LEFT
Tree showing Neanderthals as an evolutionary dead-end. A popular view on the origin of modern humans is that the Neanderthals were a geographical variant of archaic *sapiens* that became extinct when fully-modern *Homo sapiens* moved into Europe from elsewhere, possibly from Africa.

FACING PAGE
The first fully-modern human remains come from CroMagnon in France. Were CroMagnon's ancestors the Neanderthals? or, Did this modern form arise elsewhere, displacing the Neanderthals into extinction? The map shows sites from which human ancestors intermediate between *Homo erectus* and *Homo sapiens* have been found. (Numbers in parenthesis represent age: 100+ = 100000 years old.) Photographs are by courtesy of Milford Wolpoff.

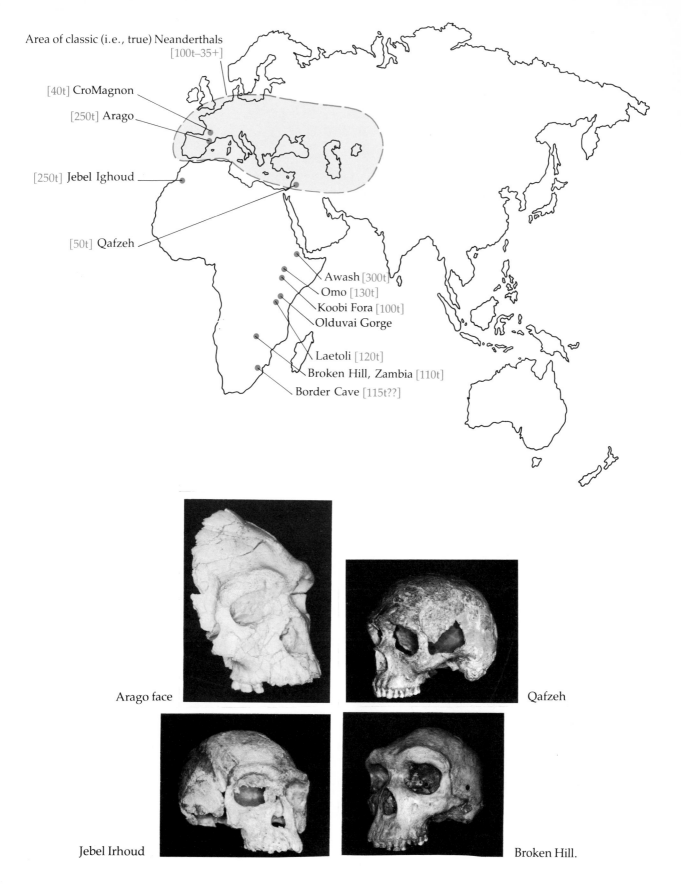

Area of classic (i.e., true) Neanderthals [100t–35+]

[40t] CroMagnon

[250t] Arago

[250t] Jebel Ighoud

[50t] Qafzeh

Awash [300t]
Omo [130t]
Koobi Fora [100t]
Olduvai Gorge

Laetoli [120t]
Broken Hill, Zambia [110t]
Border Cave [115t??]

Arago face

Qafzeh

Jebel Irhoud

Broken Hill.

Although there was a slight increase in brain size through time, the overall *erectus* pattern remained quite distinct from 1.6 million years ago to less than half a million years ago. Exactly when the transition to a more sapient form began is difficult to place in the fossil record. Some believe the transition, which involved an increase in brain size and a decrease in the robustness of skull bones, began as early as 400 000 years ago. Others prefer a much more recent date, perhaps half as long ago. Unfortunately, there are simply not enough fossils from this important period to be sure about what was happening.

There is, however, a collection of mostly incomplete crania from Europe and Africa that indicate a possible evolutionary shift to the archaic sapient form by 250 000 years ago. A skull from Petralona, Greece is a good example, dated at perhaps 300 000 to 400 000 years old. A face from Arago in the French Pyranees shows distinct sapient qualities and is dated at some 250 000 years. Parts of the skullcaps found at Swanscombe in England and Steinheim in Germany date from about the same time and also indicate a larger brain and thinner cranium. There are several others too, all of which point to a somewhat lengthy transition to the sapient grade in Europe, which presumably blossomed fully as *Homo sapiens neanderthalensis* by 100 000 years ago.

In Africa, however, there is, as yet, less evidence

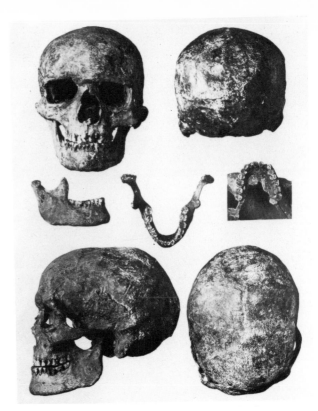

CroMagnon remains from Grotte des Enfants, France. Note the modern domed cranium. Courtesy of American Museum of Natural History.

for the emergence of the sapient grade as early as in Europe. And, of course, the classic Neanderthal form never developed there. There are, nevertheless, a series of fossils, from Tanzania, Kenya, Zambia and Ethiopia, that display a sapientization process between a little over 250 000 and 100 000 years ago. More intriguing, however, several specimens that have been somewhat uncertainly dated up to 115 000 years ago fall into the modern range of human measurements. These discoveries give a number of researchers cause to believe that modern humans arose in Africa and migrated to the rest of the Old World via the Middle East.

The oldest putative modern human specimen, part of a skullcap, comes from Border cave in South Africa. Karl Butzer, of the University of Chicago, dated the specimen at 115 000 years and he adheres to the replacement model, although he accepts there could have been interbreeding with indigenous populations. Richard Klein, also at Chicago, likewise prefers the modified replacement model, but looks to the Middle East for the first modern forms, specifically at Qafzeh in Israel, a site dated at perhaps 50 000 years old. He believes the Border cave date is

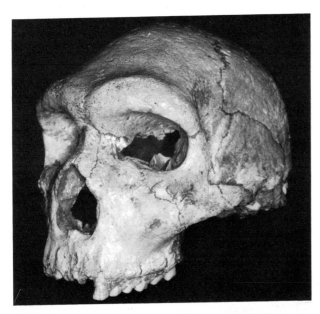

The Petralona cranium, recovered from a cave in Greece and dated between 300 000 and 400 000 years old. Although the face is partially eroded, the cranium can be seen to be a mosaic of *Homo erectus* and *Homo sapiens* features. For example, the brow ridges are ancient, while the large rounded cranium is evidence of modern development.

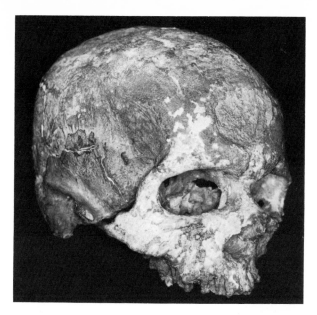

This specimen of one of the earliest *Homo sapiens sapiens* so far discovered (dated at 35 000 to 40 000 years) comes from CroMagnon, near Les Eyzies in the Dordogne, France. It had a cranial capacity of 1600 cm^3. Courtesy of Milford Wolpoff.

their departure as recorded in the fossil record should reveal something about the nature of their fate: extinction without issue, or transition to modern humans?

The picture etched in the fossil record is, however, not clearly outlined enough to be unequivocal. Although some of the Neanderthal's pronounced facial features had begun to recede in some later populations in a rather gradual manner, the disappearance of the typically massive skeletal complex was rapid. And while at some archaeological sites there is an apparently gradual shift from the Mousterian to the Upper Palaeolithic tool technologies, the inference that this change reflects a gradual transition in the people might not be validly drawn. For example, the modern forms at Qafzeh, Israel are associated with the less advanced Middle Palaeolithic tools. More to the point, perhaps, is that the modern forms, though robust, do not display any anatomical complex that would indicate a transition from Neanderthal. This conclusion, drawn by Trinkaus and fellow Harvard palaeoanthroplogist William Howells, strongly supports some form of the replacement model.

It should be noted that these modern humans of the early Upper Palaeolithic were distinctly robust as compared with populations today. They were within the range of contemporary humans, but at the upper end of the robust scale. The reduction in stockiness in the frame and in the size of teeth continued throughout the Upper Palaeolithic.

too uncertain to be accepted. The other position, that of local replacement, is strongly championed by Milford Wolpoff, of the University of Michigan. There are almost as many positions on this as there are palaeoanthropologists.

As Neanderthals represent one of the best sampled populations in prehistory, the manner of

24 / The Transition to Modern Humans: an African Study

The trajectory of human history, through to modern times, has involved an ever greater ability to extract energy from the environment. For the earliest hominids this involved being able to find and eat fruits in an environment not suited to the life style of an ape. Later, the inclusion of opportunistically scavenged meat and the digging up of underground tubers would have further widened and deepened our ancestors' economic scope, especially in the context of a cooperating, sharing social group. The development of active hunting conferred more

authorities, altered irrevocably the relationships of humans to the natural environment.

Direct evidence of a change in the economic order with the emergence of modern humans has been difficult to find, but Richard Klein, of the University of Chicago, believes he has indications from a number of coastal sites in South Africa. Klein's analysis of the types of animal bones at these sites reveals an increased efficiency in the exploitation of these natural resources. This increased efficiency might be causally related, Klein speculates, to the extinction of several large mammal species around 10 000 years ago. American researcher Paul Martin has made the same inference about the disappearance of several large species, including the giant ground sloth, in the Americas at about the same time in human history.

Through conventions of archaeology, the Middle Palaeolithic period in Europe, which coincides with

Later Stone Age artifacts

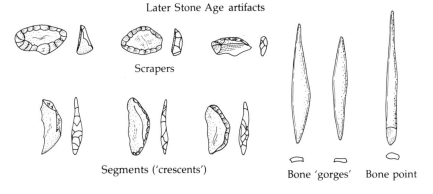

Scrapers

Segments ('crescents')

Bone 'gorges'   Bone point

Middle Stone Age artifacts

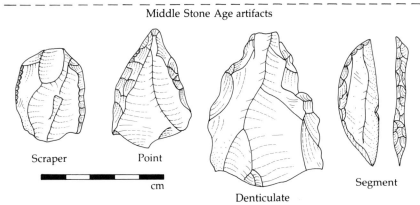

Scraper   Point

cm

Denticulate

Segment

Middle and Later Stone Age implements from a South African site. The emergence of fully-modern humans was marked by a shift from Middle Stone Age artifacts to the much smaller, finer artifacts of the Later Stone Age. Courtesy of Richard Klein.

security on the supply of energy-rich meat. Becoming an active carnivore in a world already populated by several species of successful carnivores cannot have been an easy transition. This transition probably occurred in several stages, and there is evidence that becoming an even more efficient carnivore was one factor in the emergence of *Homo sapiens sapiens*. It was a transition that, according to some

the Mousterian culture, is known in Africa as the Middle Stone Age. This covers, roughly speaking, 150 000 to around 35 000 years ago. The equivalent of the European Upper Palaeolithic is known in Africa as the Later Stone Age. The artifacts that are characteristic of the Middle and Later Stone Ages in Africa are very similar to the equivalent European periods. In both continents, the finer stone implements of the later period are accompanied by delicate bone artifacts and items of personal adornment.

The sites in which Klein was particularly interested are Klasies River Mouth cave (Middle Stone Age) and Nelson Bay cave (Middle and Later Stone Age), both of which are several hundred kilometres east of Cape Town. The people who occupied these sites had made their living by exploiting both aquatic and terrestrial resources, but the debris left behind reveal a difference in the way they went about it.

Age people were not capable of catching fish or sea birds with any proficiency, and notes that certain items of fishing gear, such as net sinkers and rudimentary hooks, are not found at Klasies whereas they are present in the Nelson Bay deposits.

Another interpretation, of course, is that the Middle Stone Age people at Klasies preferred instead to concentrate on meat on the hoof. However, again, Klein infers from the bony remains that the Later Stone Age inhabitants of Nelson Bay were more proficient hunters than their forebears. One pro-

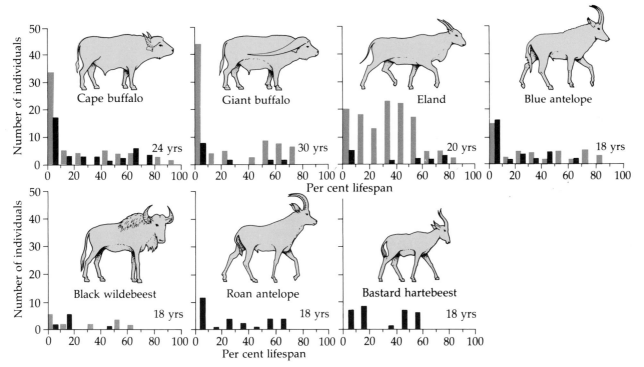

Different hunting strategies. Bars (red for Klasies River Mouth; black for Nelson Bay cave) show age distribution of prey animals. Eland and bastard hartebeest remains show a relative abundance of prime age adults. Buffalo, blue antelope and roan antelope remains are biased towards very young individuals. These two patterns indicate different hunting strategies that reflect differences in the ecology and behaviour of the prey species. Courtesy of Richard Klein.

The Klasies River Mouth cave appears to have been occupied intermittently between 130 000 and around 75 000 years ago, and then not again until relatively recently, some 5000 years ago. Among the aquatic resources exploited by the Klasies people were limpets, seals and penguins. This, incidentally, is probably the earliest evidence of systematic exploitation of marine resources so far recorded. By contrast with Klasies, the Later Stone Age people of Nelson Bay cave also killed and ate fish and flying sea birds, such as cormorants, gulls and gannets, bones of which are only rarely found in the Middle Stone Age deposits. Klein concludes that the Middle Stone

bably significant difference between animal remains in Middle as compared with Later Stone sites is the presence in the latter, but rarity in the former, of wild pigs, such as bushpigs and warthogs. These animals can be very fierce when attacked. Klein therefore believes that the Later Stone Age people had developed methods for dealing with dangerous prey whereas the earlier inhabitants had not. Although the use of bows and arrows by the Nelson Bay inhabitants would be an obvious explanation of this difference, so far no archaeological evidence has been found to support this idea.

The great proportion of prey taken by both Middle

and Stone Age hunters was bovine, that is various types of antelope and buffalos. For the Middle Stone Age people, eland and bastard hartebeest were apparently particularly important prey species, although Cape buffalo, giant buffalo, blue antelope, black wildebeest and roan antelope were also killed or scavenged. The same range of prey occurs in the Later Stone Age site but, proportionately, eland and bastard hartebeest are much less important there. Klein notes that eland and bastard hartebeest are particularly docile animals, and argues that the relatively poor hunting skills of the Klasies hunters forced them to concentrate on these creatures as prey. Because the bones of prey animals in the Nelson Bay cave more nearly reflect the live animal populations at the time, Klein concludes that the Later Stone Age hunters had become more proficient.

The individual prey animals in both cave sites were mainly either very young or rather old, which reflects very much the age pattern of carnivore kills, as these animals are, for various reasons, the most vulner-able. For the eland and the bastard hartebeest, however, the pattern was different: the age distribution of the bones at the sites was very much what one would expect in a typical herd. Klein therefore suspects that the stone age hunters were able to drive these prey into traps or over falls. (The ease with which eland are driven is well known and indeed was noted by Theodore Roosevelt in his *African Game Trails*.)

Eland herds are generally very widely dispersed, and so even proficient hunters would be unlikely to make a great impact on the population as a whole. But, says Klein, the evidently greater skill of the Later Stone Age hunters may have become a threat to other species, which might account for the extinction of several large mammal species 12 000 to 10 000 years ago, an event that cannot be accounted for by environmental changes alone, he suggests. This, in addition to Martin's equivalent suggestion for certain species in the Americas, perhaps signals the first major destructive impact of modern humans on the world in which they live.

25/Brain Growth and Intelligence

A major question about hominid history is why we became so very intelligent, particularly over the past two million years. The primate order as a whole is the most generously endowed of all animal orders as far as brain capacity relative to body size is concerned. And within the order there is, in general, an ascending scale of relative brain capacity that runs from the prosimians through the monkeys to the apes. The human brain is three times as big as an ape's would be if it had the same body size.

Increase in brain size is a persistent theme of evolutionary history in two particular respects. First, the progression through more and more advanced animal groups—from amphibians through reptiles to mammals—is marked at each step by a substantial

The second pattern of brain size increase in evolution is seen within individual lineages through time. This, typically, is associated with a parallel tendency for an increase in body size through evolutionary time. Some of the growth in brain size from around 400 cm^3 in the earliest hominids to an average of 1360 cm^3 today can be accounted for in terms of an increase in the size of the body, but most cannot. There was a real and dramatic enhancement of encephalization in hominids that is not matched, nor even approached, by any other animal lineage.

At least as important as brain size is the overall organization of the brain, specifically the relative proportions of the various lobes. Here, the work of Ralph Holloway at Columbia University, New York has produced something of a surprise with respect to hominid brains when compared with apes.

The occipital lobe, located at the back of the brain, is involved with visual functions whereas the temporal lobe, at the side, is responsible for memory. Sensory integration and association resides in the parietal lobe, which is located on the top of the brain above the temporal lobe. And some aspects of motor

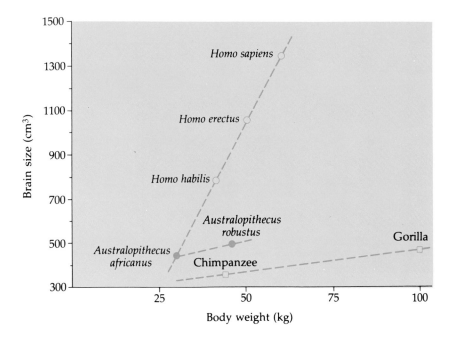

Brain size and body weight in humans and apes. The steep slope of the brain/body weight graph in humans as compared with apes reveals a significantly greater encephalization.

leap in the degree of encephalization displayed by each group as a whole. These stepwise mental increments between the major animal classes reflect gestalt jumps in the complexity of neural processing involved in the animal's daily lives. Each increment has been accompanied by an ever greater learning capacity as opposed to genetically determined fixed action patterns.

control and emotional behaviour are performed by the frontal lobe. All four lobes are, of course, present in both right and left hemispheres of both humans and apes, however the size emphasis is different. The human pattern, for example, emphasizes the frontal, temporal and parietal lobes, while the occipital lobe is quite small. In the ape pattern the frontal, temporal and parietal lobes are relatively

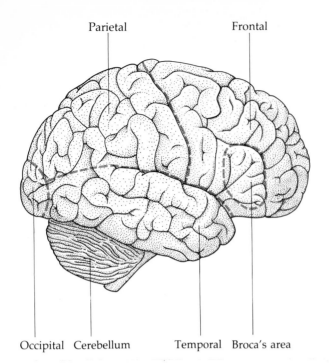

Parietal   Frontal

Occipital   Cerebellum   Temporal   Broca's area

Diagram of the typical ape and human brain pattern. The large human brain (left) compared with that of a chimpanzee is also distinguished by its relatively small

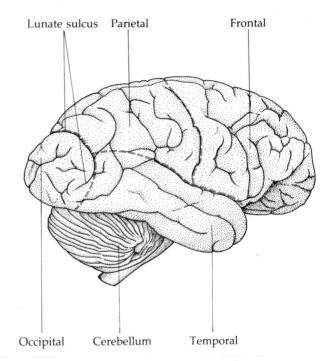

Lunate sulcus   Parietal   Frontal

Occipital   Cerebellum   Temporal

occipital lobe and large parietal lobe. From *The Casts of Fossil Hominid Brains* by Ralph Holloway. Copyright © (1974) by Scientific American, Inc. All rights reserved.

small while the occipital lobe is emphasized. These contrasting organizational patterns, in addition to the overall difference in size, are presumably responsible for the greater intelligence and more advanced technological and social skills of humans.

Holloway has for more than a decade been examining the brain organization of early hominids, either by taking advantage of natural endocasts that sometimes form inside fossil crania or by making a latex mould of the inner surface of the cranium. Although details are few, the interior surface of the cranium does bear a shadowy signature of the general shape of the brain that it once housed. This has been sufficient to determine whether or not fossil hominids had typically human brains, typically ape brains, or something in between.

Although a very reasonable guess would have been to choose the last of these three possibilities, it would have been wrong, according to Holloway. In all the fossil hominid skulls he has so far analysed, and this includes the frustratingly fragmentary material from the oldest specimens at the Hadar, the brain organization has been found to be typically human. It is difficult to escape the conclusion that even from the very beginning of the hominid lineage, there was a marked divergence in behaviour

patterns as compared with those in the apes. This difference in brain organization was almost certainly not simply a shift in the underlying mechanics of locomotion, from quadrapedalism to upright walking, and it cannot be associated with stone toolmaking, which did not start until several million years after the hominid lineage arose. It should be noted that in more recent analysis Dean Falk, of the University of Puerto Rico, finds that brain reorganization becomes significant only in *Homo*, not in *Australopithecus*. Whichever interpretation is correct, the question remains, was there a social behaviour pattern specifically associated with hominids which was served by a particular brain organization?

The world that any animal inhabits is, perforce, the one created in its brain by the integration of sensory inputs, such as sight, sound, smell and touch. The more sophisticated the inputs and their neural processing, the more complex and more real will be the inner world built by the brain. Such an increase in complexity is what underlies the incremental increase in encephalization from amphibians through reptiles to mammals. One can reasonably infer that life for the average mammal is less predictable and more demanding than for the average amphibian or reptile. For primates, especially humans, this

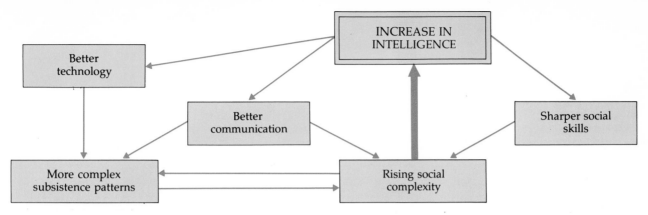

Social complexity and increased intelligence. The need to cope with rising social complexity—including more and more demanding subsistence patterns but, particularly, a more ramified social structure and unpredictable social interactions—may have been a keen selection pressure for increased intelligence.

observation is even more pertinent.

When British psychologist Nicholas Humphrey was contemplating the question of what it is that is so complex and unpredictable in the lives of humans that they need to be so intelligent, he came up with a surprising answer. It is not the world of practical affairs that is so difficult to master, it is the behaviour of other members of one's species. Moreover, Holloway argues, the extensive ramifications of social interaction were likely to have been a powerful force in the evolution of our very large brains.

Most primates are intensely social, particularly the higher primates and the argument about intelligence and the complexity of social interactions applies here too. Why, for instance, should chimpanzees be mentally equipped to solve intellectually challenging puzzles that human researchers delight in presenting them? The chimpanzee's subsistence does not seem so very keen a task that it would require so honed an intellect, says Humphrey. But its social life is unquestionably complex.

For humans, the adoption of a wider dietary base within a food-sharing social group would have placed even sharper demands on the ability to deal with the complex and unpredictable. The great technological skills and fine artistic expression of humans might, therefore, be the by-product of the need to be socially adept.

26/The Origins of Language

Only humans posess a spoken propositional language, and it is so powerful a facility that one must reasonably suppose that it played an important part in the species' evolutionary history. Two major questions can be asked of human language in an evolutionary context. First, When did it emerge? And second, Under what selection pressure did it develop? To neither question is there a definitive answer, perhaps because of the evanescent nature of language, but there are some informed speculations.

Considerable neural machinery underlies language abilities, which must account for some of the increase in brain size discussed in the previous section. In addition, in most people language centres are located principally within the left cerebral hemisphere, which, as a consequence, is slightly larger than the right. Two centres, Wernicke's area at the side of the brain and Broca's area towards the front, are responsible for the structure and sense of human speech and the coordination of the relevant throat and mouth muscles. Broca's area, and to a lesser extent Wernicke's area, are denoted by slight swellings on the surfaces of the left hemisphere.

Unfortunately, the laterality of hemispheric dominance and the swelling over Broca's area cannot be taken as clearly diagnostic of language abilities because in recent years apes have been discovered to possess similar, if less pronounced, anatomical

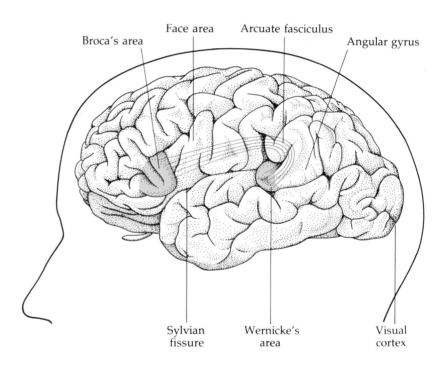

Language centres. Wernicke's area, which appears to be responsible for content and comprehension of speech, is connected by a nerve bundle called the arcuate fasciculus to Broca's area, which controls the muscles of the lips, jaw, tongue, soft palate and vocal cords during speech. These language centres are usually located in the left cerebral hemisphere, including many left-handed people. From *Language and the Brain* by Norman Geschwind. Copyright © (1972) by Scientific American, Inc. All rights reserved.

Because the products of a language facility—words—do not enter the fossil record one is forced to deal indirectly with the question of when the facility arose. One possibility is to look to the brain itself, or rather to fossil endocasts, and to the brain case, specifically the base of the cranium, which carries some clues as to the organization of the pharynx, or voice box. Another, is to scrutinize other hominid products, such as tools and art objects, for some indication of the type of cognitive capabilities necessary for spoken language.

features. Where it has been possible to look for language correlates in endocasts, with the famous 1470 skull from Lake Turkana for example, the hemispheric dominance and the Broca's swelling can be detected, which is perhaps neither very surprising nor, unfortunately, very helpful.

The underside of the modern human cranium is flexed, or vaulted, apparently to accommodate the modified anatomy associated with the vocal apparatus. The basi-cranium in apes is much flatter. Jeffrey Laitman, of Mount Sinai School of Medicine, New

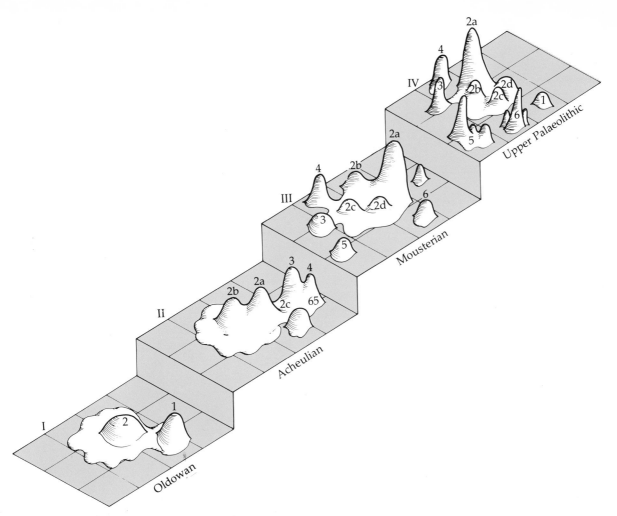

Sharpening the mind: sharpening the tongue. With the passage of time and the emergence of new species along the *Homo* lineage, stone tool-making became even more systematic and orderly. Peaks in the diagram represent identifiable artifact modes, with tall, narrow peaks implying highly standardized products. The increased orderliness in stone tool manufacture must, argues archaeologist Glynn Isaac, reflect a more and more ordered set of cognitive processes that eventually involved spoken language. I, eg Oldowan: 1 = core-choppers, 2 = casual scrapers; II, eg Acheulian (Olorgesailie): 2a = scrapers, 2b = nosed scrapers, 2c = large scrapers, 3 = handaxes, 4 = cleavers, 5 = picks, 6 $\doteq$ discoids; III, eg Mousterian: 2a = racloir, 2b = grattoir, 2c = r. convergent, 3 = percoir, 4 = point, 5 = burin, 6 = biface; IV, Upper Palaeolithic: 2a = grattoir, 2b = nosed scraper, 2c = raclette, etc., 3 = percoir, 4 = point, 5 = burins, 6 = backed blades, etc. Courtesy of Glynn Isaac.

York, has examined these landmarks of language in early hominids and finds an ape-like pattern in the australopithecines and an emerging, but not complete, human-like arrangement in *Homo erectus*. Unfortunately there are no suitable basi-crania of *Homo habilis* on which to make a reliable judgement.

Turning now to non-language products of ancient hominid brains, archaeologist Glynn Isaac has analysed the progression of stone tool-making over the past 2.5 million years and suggests there are inferences to be drawn. Isaac notes that the whole sequence is remarkable not for the innovation of new implements but for the increasing standardization of types through time. It is, he suggests, as if some kind of rule system is being imposed, with incremental leaps in its importance at 1.5 million, 150 000 and 40 000 years ago (see sections 20 and 21). These dates coincide with the appearance of *Homo erectus*, archaic *Homo sapiens* and *Homo sapiens sapiens*. The standardization of tool manufacture reveals as much about the nature of the societies in which tools are made as it does about the practical requirements of such implements, suggests Isaac. The progressive orderliness seen in stone tool-kits is just one ex-

Dated at 300 000 years, this engraved rib (full picture, detail and drawing) from Peche de L'Aze in France, is one of the earliest examples of abstract symbolism in the prehistoric record. Abstract symbolism must betoken some facility for spoken language. Courtesy of Alexander Marshack.

pression of an increasingly ordered society. And such structuring of society—in terms of social relationships, of the manner in which resources are exploited, and expression of group identity—is difficult to conceive of without the eventual emergence of a complex spoken language.

Ralph Holloway links spoken language and tool-making even more closely together, arguing that the cognitive processes underlying each are very similar. Both processes involve the sequential elaboration of component parts that, if inserted out of a prescribed order, make nonsense of the final product. Gordon Hewes, an anthropologist at the University of Colorado, concurs with the suggestion of an analogy between speech and tool-making and adds the observation that tongue and mouth movements are commonly associated with precise manual tasks. Hewes is a proponent of the idea that hand gestures probably predated the emergence of spoken language as a form of communication. As in all animals, non-verbal communication must have been an important part of early hominid interaction; and, of course, still is.

The second group of non-language products of the

prehuman brain, that is, art objects, are clearly out of the utilitarian realm and directly within some kind of context of symbolism. 'What use is a symbolic object in the absence of an abstract cultural context?' asks Alexander Marshack of New York University. One can envisage such activities only where the cultural context has been created through spoken language.

The oldest abstract artifact known so far is a 300 000 year old ox rib from France on which is carved a series of connected, festooned double arcs. This type of pattern is common in the Upper Palaeolithic, from 40 000 years on, but between these two dates there is virtually nothing comparable. Of equal age to the engraved ox rib, is the presence of red ochre within a coastal springtime shelter that has been excavated in Southern France. Was this used for ritualistic body decoration, an activity that must be one of the most vivid and yet least archaeologically visible expressions of the transcendental human spirit?

From Mousterian times there is a growing number of examples of engraved bone and ivory that indicate an awareness of and an ability to deal with the abstract, as does the list of ritualistic burials. Once the Upper Palaeolithic begins, however, the numbers and elaboration of such activities give a sense that some kind of threshold has been passed. This threshold may well have been an incremental leap in the facility for spoken language, not the first innovation of this powerful mode of communication but perhaps a substantial refinement of it.

There can be no doubt about the benefits of a spoken language as a means of communication, and these may have been a part at least of the selection pressure that produced it. However, as Harry Jerison speculates, the evolution of language may also have been a response to the ever expanding need to create an improved model of the world in one's brain. Jerison, a neurobiologist at the University of California at Los Angeles, says: 'I am proposing . . .

that the role of language in communication first evolved as a side-effect of its basic role in the construction of reality.' Language is, he says, a means of constructing mental imagery. 'We need language more to tell stories than to direct actions.'

As with the evolution of intelligence, it is useful to contemplate the emergence of language very specifically within the social context of an economically interdependent society. It is not that the world of practical affairs is unimportant in the progress of human prehistory, but that the prime context of such affairs is that of complex social interactions. 'In hearing or reading another's words we literally share another's consciousness, and it is that familiar use of language that is unique to man', says Jerison.

27/Art in Prehistory

Modern humans arose at the midpoint of the last Ice Age, which began about 75 000 years ago and ended 10 000 years ago. The ice epoch was at its most severe towards its end, some 18 000 years ago, a point which coincides with the apogee of prehistoric art. In many ways, prehistoric art is the art of the Ice Age and especially of Ice Age Europe where favourable conditions of preservation have protected against the ravages of time the heart-stopping paintings deep in limestone caves.

Although most general interest in prehistoric art has centred on colourful images painted on cave walls and rock shelters, many thousands of carved and engraved pieces of bone and ivory have been discovered from the same period that, while not as immediately impressive as the painting, are at least as intricate in their crafting and as engaging in their potential meaning. A comparison of the two forms of art shows, to some, that the paintings were probably the focus of some form of community activity while the portable art objects, art mobilier, were probably more personal items. Beyond this general distinction there are differences between the paintings and the art mobilier that form one of the more intriguing studies of Ice Age art.

The first painted caves in Europe were discovered towards the end of the last century, and now there are almost 200 such sites known, principally from Southern France and Northern Spain. The first systematic studies of European wall art were carried out by the great French prehistorian, Abbé Henri Breuil. He carefully copied many of the pictures, attempted a chronology of the sites, and finally concluded that the images were part of the Ice Age people's hunting magic.

The great majority of images are indeed of contemporary animals while paintings of people are virtually non-existent. And where human form does appear, it is mostly in crude outline or in stick-like charicature which contrasts with the fine and acute detail often displayed in the animal images. The collection of images in the great majority of caves

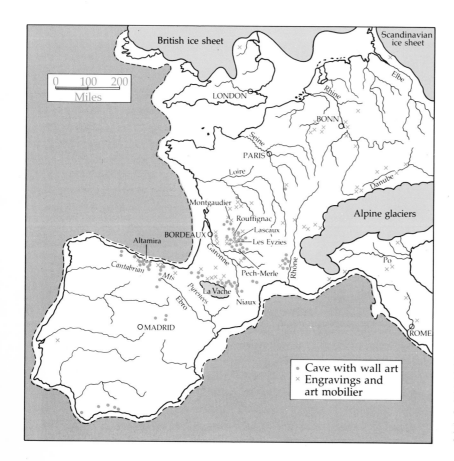

Distribution of art sites in Europe. The limestone caves of Ice Age Europe have preserved a rich legacy of Palaeolithic art. Although there was a certain stylistic continuity in cave painting, motifs in art mobilier displayed much more variability.

(a)

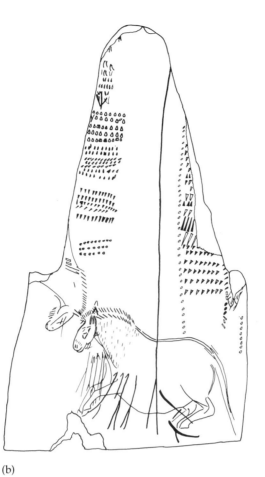

(b)

Examples of cave art.

(a) Fragment of reindeer antler from La Marche, France, around 12 000 years old. Apparently used as an implement for shaping flint tools, the antler fragment is engraved with a pregnant mare, which seems to have been symbolically killed by a series of engraved arrows. Above the horse is a set of notches made at different times by different tools. Marshack interprets the marks as a notation series, perhaps documenting the passing lunar cycles.

(b) A drawing of the engraving 'unrolled'.

Below

(e) Vogelherd horse, carved from mammoth ivory about 30 000 years ago and worn smooth by frequent handling over a long period of time. The horse, which is the oldest animal carving known, measures 5 cm.

(f) The black outline of this horse was painted on the wall of a cave, Peche-Merle, in France 15 000 years ago. Infrared analysis indicates that the mixture of black and red dots were added over a period of time. The black hand stencils are also later additions. Does the Peche-Merle horse, one of two in the cave, indicate 'use' of art? Courtesty of Alexander Marshack.

(c)

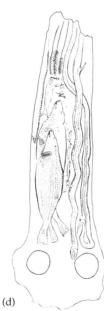

(d)

(c) Engraved antler baton from Montgaudier, France, dated about 10 000 years old. Perhaps used in straightening the shafts of arrows or even spears, the collection of items engraved on the baton suggests a representation of Spring.

(d) A drawing of the engraved antler baton 'unrolled'.

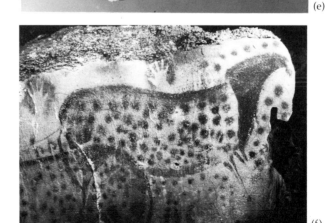

(e)

(f)

gives the impression of having been built up piece-meal. Each picture seems to be the product of a discrete event, often showing little regard for content or position of existing pictures. This, together with the almost universal absence in the painted caves of debris from day to day habitation, encouraged the view that the paintings were performed in a ritual-istic, sacrosanct setting. Hunting magic was an obvious interpretation, and it held sway from the early decades of this century until the 1960s.

There were, however, several problems with the hunting magic notion. In many areas where related archaeological remains showed reindeer to be the principal source of meat, images of other animals, usually bison, were much more important in the paintings. And only a tiny minority of painted animals show any sign of ritualistic killing.

During the 1960s two French prehistorians, André Leroi-Gourhan and Annette Laming-Emperaire de-veloped another interpretation, which emphasized the social context of the cave art. They noted that the inventory of animals depicted was comparable throughout Europe and that the presentation was remarkably stable through time; an observation that, incidentally, contrasts with the much more locally idiosyncratic nature of art mobilier. Cave art, for Leroi-Gourhan and Laming-Emperaire, reflected the duality of maleness and femaleness in society. Certain animal images were said to represent maleness while others were female. And the arrangement of images in a cave might put the females at the centre with the males around the periphery, thereby reflecting a certain type of social structure. Even though the two prehistorians did not agree on which class of images was meant to represent which sex, their thesis encouraged a greater emphasis on social context in the inter-pretation of Ice Age art.

In more recent times, there has been a more explicit recognition that Ice Age art encompassed many media and that no single explanation can be applied to all expressions. An interesting example of the coming together of different forms of art is Altamira cave in Spain; Altamira and Lascaux in South-west France represent the most richly decorated sites yet known. A circle of bison, sur-rounded by a number of other different animals, form the centrepiece of the ceiling at Altamira. According to evidence from palaeoecology, and from the discovery there of a range of tool types and art mobilier objects distinctive of several different neighbouring locales, Altamira appears to have been a site of congregation, probably in the Autumn, for many different social groups. Margaret Conkey, of the State University of New York, suggests that the seasonal aggregation of social groups is reflected in the aggregation of bison on the cave ceiling. The painting, in other words, is an encapsulation of a major social and economic activity.

Many art mobilier objects are decorated with geometric patterns; some have pictures of animals, fish and plants; and others have series of seemingly random notches. Alexander Marshack's detailed studies of such objects reveals something more coherent to this aspect of prehistoric art than has previously been supposed. For example, he inter-prets the images of a male and female seal, a male salmon, two coiled snakes and a flower in bloom, all engraved on a baton made from reindeer antler, as a representation of Spring. Other apparently jumbled images engraved on ivory knives, flaking tools and the like can be similarly interpreted as seasonal vignettes.

One of Marshack's more revolutionary suggestions is the notion of repeated use of Ice Age art. Not only do some 30 000 year-old ivory carved horse figures from Vogelherd in Germany show smoothing from continued handling, for example, but certain items bear series of marks that microscopic analysis shows to have been applied at different times. Marshack speculates that some of these might even have been notations of lunar cycles, for example. His studies of some cave paintings, using infrared light, also reveal repeated addition and modification to images, which is another facet of art in use.

It is difficult to know, at this distance, what was in the minds of the palaeolithic artists. Why, for ex-ample, is the human form so rare? What was the true meaning of the crude Venus-like figures so common in Northern Europe? What signal was meant by the world-wide practice of creating a stencil print of the hand.

Art did not stop with the passing of the Ice Age; however it did change in form. Geometric patterns became a much more important component of paintings. The human form became more common too, often depicted in combat with others, something that was rare before the advent of the agricultural revolution 10 000 years ago.

28/Expanding Populations

There have been two major population dispersals in human prehistory. The first involved the move from the tropics, where hominids evolved, into more temperate regions to the north. This occurred around one million years ago and involved populations of *Homo erectus.* The second dispersal was of *Homo sapiens sapiens,* to the New World via the north and Australia in the south. The dates for these most recent migrations are yet to be settled, but they both might well have been very soon after the origin of fully modern humans.

There has often been a tendency to contemplate aspects of human history in isolation from the history of other animal groups. Indeed, there must be some respects in which the course of human history has been determined solely by the rather special be-

spirit's urge for new lands, our ancestors were simply tracking their subsistence potential through new prey populations, as were other predators.

The arrival of early humans in North America, and their subsequent migration to the southern continent, has long been a matter of intense debate. Some people argue that the Americas were unpopulated by humans prior to 12 000 years ago, while others consider an arrival date of 35 000 to 40 000 years ago more likely.

The route of migration to North America was, of course, through Siberia and Alaska, via the Bering Land Bridge that was periodically exposed as vast quantities of water were soaked up into the expanded ice caps of the last Ice Age. Major ice advances reduced sea levels by up to 100 m, which caused North-east Asia and North-west America to be joined in a 1500 km-wide plain known as Beringia. According to some authorities, glaciation was severe enough to expose all or part of Beringia between 40 000 and 35 000 years ago, again between 25 000 and 14 000 years ago, and at minor intervals between 14 000 and 10 000 years ago.

Clearly, the exposure of Beringia during glacial

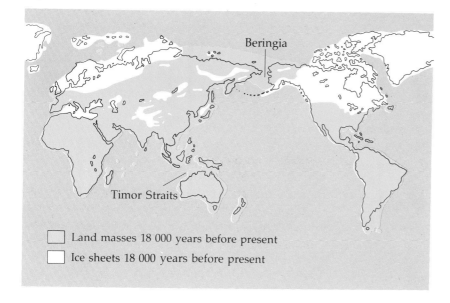

Land masses 18 000 years before present
Ice sheets 18 000 years before present

Migration routes to Australia and America. Eighteen thousand years before present was the apogee of the last glaciation (75 000–10 000 BP). Expanded glacial cover (white areas) lowered sea levels to expose shallow continental shelf (shaded areas over current coastlines). Although glaciation was less than shown 40 000 years ago, the Timor Straits were still considerably narrowed, facilitating the migration into Australia (and later into Tasmania). The reduced glaciation some 20 000 to 30 000 years ago might also have left an ice-free corridor linking North America and Siberia.

havioural repertoire displayed by the genus *Homo.* Equally, however, the human lineage on occasions must have responded to ecological changes in ways parallel to that in other animals. For example, Alan Turner, of the Transvaal Museum in Pretoria, argues that the initial dispersal from Africa and the later migration to North America can be viewed as territorial expansions in concert with other large predators. Rather than answering some inward

advances sets certain limits on migration possibilities into North America. These limits might have been further constrained by an ice barrier formed by the joining of the Laurentide ice mass centred around the Hudson Bay region and the Cordilleran ice mass centred on the Rockies and the coastal mountains. It has been argued that the coalesced ice masses formed an impenetrable barrier between 25 000 and 9000 years ago. There is, however, a steadily more popular view that an ice-free corridor existed between the ice masses throughout the Ice Age, with the possible exception of the peak of glaciation around 18 000 years ago.

Beringia would have been a virtually treeless tundra, rich in grasses and sage, upon which grazing animals such as mammoth, horse, bison, elk and the Siberian steppe antelope thrived. Wolves, lions and humans would have found sufficient prey on which to subsist.

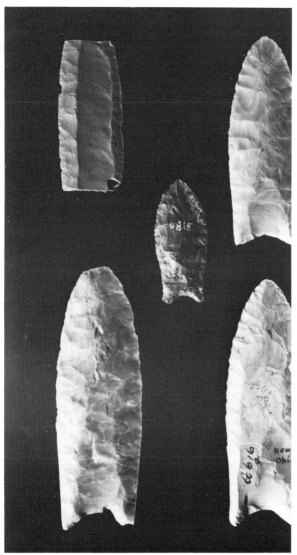

Clovis projectile points. Courtesy of American Museum of Natural History.

There is no dispute that humans had arrived in the Americas by 11 000 years ago. The distribution of a very characteristic artifact, the fluted point, in well-dated sites from the Pacific to the Atlantic coast in North America, and from Alaska to Central Mexico, confirms this. The fluted point, also known as the Clovis point, had spread through the continents with great rapidity. The question is, Does the rapid dispersal of the Clovis culture reflect an expeditious entry into a land previously unoccupied by humans? Or, Was this, in part at least, a diffusion of a new culture among existing populations?

There are a number of archaeological sites throughout the Americas that are claimed to predate the Clovis culture. A major contender is from the Old Crow Basin in the Yukon Territory of Alaska. Canadian researchers have unearthed what they believe to be bone implements dated at 27 000 years. Some putative implements from the area may well be at least 10 000 years older still. Another putative preClovis site is located in the Peruvian highlands. American archaeologist Richard MacNeish has excavated several occupation layers there in Pikimachay cave which, he says, date back to between 25 000 and 21 000 years ago.

If either the Yukon or the Peruvian site is validated for its age, then preClovis occupation of the Americas is settled. Meanwhile, the long list of other putative early sites begs serious consideration for this conclusion. Human remains from early sites are, unfortunately, rather rare. And the one site that is well blessed with human material has been the focus of a dating dispute for many years. A group of 11 human skeletons was found at Del Mar and Sunnyvale in Southern California. Originally dated at between 48 000 and 70 000 years, these fossils have recently been more reliably placed at between 11 000 and 8300 years.

A very reasonable conclusion on the peopling of the Americas is that it began at least 35 000 years ago, but may well have included waves of immigrants at later dates too. The Clovis culture, so typical of the most recent 11 000 years, was almost certainly an American invention rather than an import from Asia.

Migration to Australia, by contrast with America, must have involved a sea journey. Although the lowered sea levels of the Ice Age would have exposed more land than is above water now, Australian immigrants would have had to island hop across the Timor Strait. Sea trips of 19, 29 and 87 km would have been necessary to reach this southern continent.

The earliest human remains in Australia are on the shores of an extinct lake in New South Wales. This Lake Mungo site is firmly dated at 32 750 years, which clearly sets a minimum date for occupation.

This caribou tibia, which has been notched at one end, is the oldest putative artifact in North America. It was found at Old Crow, Yukon Territory, and is dated at about 27 000 years. The tool was apparently used to deflesh skins. Courtesy of National Museums of Canada.

There are claims for earlier sites, also in New South Wales, and these are put at 50 000 years, but are still tentative.

Settlers in this southern continent faced conditions as harsh as those populations in Northern Eurasia. There are sites in Tasmania, for example, dated at over 20 000 years ago, that were virtually within sight of the southern ice cap. *Homo sapiens sapiens* had indeed displayed astonishing adaptive capabilities in its virtually complete occupation of the globe.

One curious aspect of the occupation of Australia is that, unlike virtually every other region of the world, the inhabitants did not shift from hunting and gathering to some form of agriculture. Agriculture came to the continent only in the late eighteenth century, with the arrival of the European settlers.

29/The Agricultural Revolution

The traditional view of the Agricultural Revolution is that at the end of the last Ice Age, 10000 years ago, there was a dramatic global shift in human subsistence patterns from nomadic hunting and gathering to sedentary food production. A world that at the point of the revolution supported perhaps just five to ten million people was set on a trajectory to overcrowding: within 8000 years the population was already 300 million, and growing fast.

There is no doubt that the change in subsistence patterns encompassed by the Agricultural Revolution was dramatic by any standards, but the development of food production and its eventual

major centres have been identified, from which, according to genetic studies, people migrated taking their new form of subsistence with them. The first is the classic centre known as the Fertile Crescent, which was an arc of fertile land running through Israel, Jordan, Syria, Turkey and Iran: agriculture was established here by 10000 years ago. A second centre, established 7000 years ago, was in China. Slightly later still, around 5000 years ago, the New World began to adopt sedentary food production, centred in Mesoamerica.

The question of why food production should replace the demonstrably secure and carefree life of hunting and gathering is a matter of some debate. Since it was a global phenomenon arising independently in many areas, prehistorians have tended to look for global answers. The most obvious candidate is the climate, as the rise of agriculture, in its earliest stages at least, coincided with the relatively abrupt end of the Ice Age. Gordon Hillman, of University College, Cardiff, argues that the warmer climes

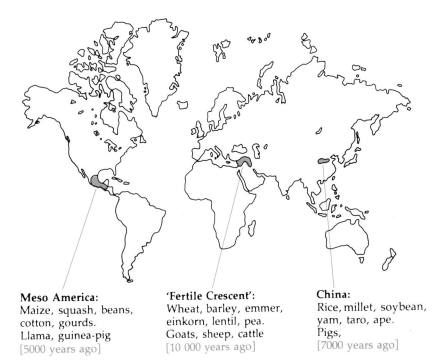

Meso America:
Maize, squash, beans, cotton, gourds.
Llama, guinea-pig
[5000 years ago]

'Fertile Crescent':
Wheat, barley, emmer, einkorn, lentil, pea.
Goats, sheep, cattle
[10 000 years ago]

China:
Rice, millet, soybean, yam, taro, ape.
Pigs,
[7000 years ago]

The three major centres of agricultural development.

dominance of economic life was probably drawn out over a longer period of time than has been generally supposed. There are, for example, tantalizing clues suggesting a certain degree of control over food resources as long as 30000 years ago.

One of the interesting features of the adoption of agriculture was its apparently independent invention in several different parts of the globe. Three

would have caused the spread of wild grasses from sheltered woodland to the open steppe of the Middle East. These grasses formed the basis of early agriculture in the region, and would have been available in great abundance.

American anthropologist Richard MacNeish agrees that environmental changes caused by postglacial climates were probably important in the onset of systematic food production in the Americas too. He has analysed the changes in vegetation in the Tehuacan Valley of Mexico and concludes that its inhabitants would have been forced to change their previous hunting and gathering activities, possibly to include crop cultivation.

The postglacial period must indeed have brought substantial environmental changes, but, argues Mark Cohen of the State University of New York, similar changes had occurred in earlier times, so what was special about the shift 10 000 years ago? Population pressure is his answer. Population growth over the millenia, since the emergence of

University of Michigan, are in agreement with her. Flannery says that the populations in Mexico when agriculture was taking root there were simply not high enough to have caused food shortages through traditional hunting and gathering.

Bender, developing certain ideas put forward in the 1960s by American prehistorian Robert Braidwood, adduces cultural rather than external pressures as the spur to a new mode of food-getting. Hunting and gathering societies were becoming more and more complex and hierarchical, she says, as indicated by the appearance of trade items and status burials long before the Agricultural Revolution

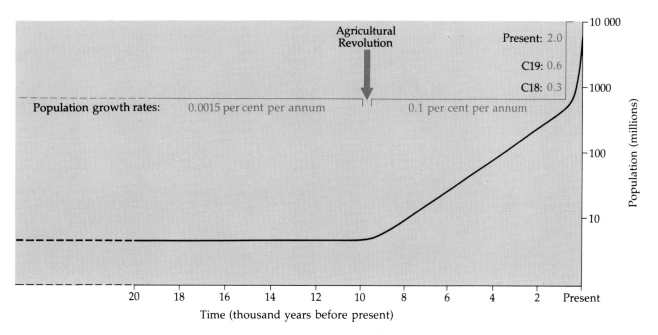

Population change since the Neolithic. Intensive food production and the consequent increase in population

densities associated with the Agricultural Revolution ignited a population explosion.

modern humans, had eventually reached a point at which local groups were finding it difficult to obtain sufficient food from hunting and gathering, which requires people to live in groups of about 25 individuals and each group to have access to at least 75 km². As a result, people began to nurture their own food crops and control livestock, thus allowing a higher population density in any given area.

Although no one denies that higher population densities were a consequence of systematic food production, not everyone agrees that population growth was a cause of the switch to agriculture. Barbara Bender, a British anthropologist, insists that the evidence simply does not support Cohen's argument. Others, including Kent Flannery, of the

occurred. The elaboration of alliances and trade between neighbouring groups would have generated pressure to produce more and more surplus goods, which eventually would have been most efficiently served by a sedentary economy. There is no doubt that trading and political alliances very quickly became very important between villages, towns and cities based on agriculture, with Jericho being a prime example. Whether such activities were a thread of continuity with earlier economies, and a causative factor in the emergence of the new economic order, remains to be established.

The history of the Oaxaca Valley in the southern highlands of Mexico illustrates what was probably a general pattern in the adoption of agriculture and its

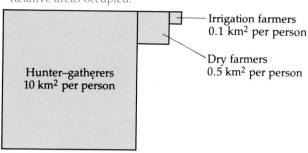

Relative areas occupied:

Hunter–gatherers
10 km² per person

Irrigation farmers
0.1 km² per person

Dry farmers
0.5 km² per person

Relative areas occupied by hunters and gatherers compared with farmers. Intensive food production and the sedentary mode of habitation among farmers greatly decreases the land area required per person compared with nomadic hunting and gathering.

consequences. Between 10 000 and 5000 years ago the people there gradually supplemented their hunting and gathering with opportunistic horti-culture of squash, corn and gourds. However, as soon as there was a complete commitment to maize agriculture the pace of change shifted dramatically. Within 1000 years the valley was transformed from a sparse scattering of village settlements to a major urban complex dominated by the artificially elevated city of Monte Alban, which covered 175 acres.

Elaborate architecture and carvings are evidence of extensive trade and political connections and reveal the use of brutal force in subjugating those who did not readily submit to the power of Monte Alban.

Prior to 10 000 years ago the evidence for control of food resources is scattered but persuasive. Extensive cultivation of wild barley has been detected in 18 000 year old sites in the Nile Valley, for example. Cattle were clearly domesticated well before 13 000 years ago in East Africa. A 15 000 year-old engraving of a horse's head decked in a form of harness from France indicates a degree of control over wild animals not generally acknowledged, as does evidence of crib biting in horses at another French site, but twice as old.

Paul Bahn, an English archaeologist, sees further evidence of control over resources in certain cave sites whose occupants specialized in particular prey animals, such as ibex and reindeer. If Upper Palaeo-lithic people were indeed managing wild game so as to control their food resources, it is likely too that they were engaging in opportunistic horticulture. The fuse for the Agricultural Revolution may have smouldered for many millenia before finally igniting the explosion.

30/Culture and the Human Experience

Darwinian evolution concerned the nature of genetic change through time, and specifically the differential survival of the fittest genotypes. Organisms passively adapt to changing ecological circumstances through a slow, generation by generation change in their gene complexes, through natural selection. Humans, by contrast, are likely to respond to such challenges by changing their behaviour or by changing the environment. Human experience is now quintessentially within the realm of culture, not in raw biology. Change, which once was measured in hundreds of millenia, has been ever more telescoped in human experience and is now measured in hundreds of days.

Again, in contrast with the evolutionary trajectory of most organisms, humans have become one of the

being human. As American anthropologist Clifford Geertz has observed, 'Man's great capacity for learning, his plasticity, has often been remarked upon, but what is even more critical is his extreme dependence upon a certain sort of learning: the attainment of concepts, the apprehension and application of systems of social meaning.' He goes on to say, 'Men without culture would not be the clever savages of Golding's *Lord of the Flies* thrown back upon the cruel wisdom of their animal instincts; nor would they be the nature's noblemen of enlightenment primitivism, or, even, as classical anthropological theory would imply, intrinsically talented apes who had somehow failed to find themselves. They would be unworkable monstrosities with very few instincts, fewer recognizable sentiments, and no intellect: mental basket cases.'

Humans produce culture, no one doubts; but, in a very real sense, culture produces humans, so Geertz contends. He does not suggest that every aspect of human behaviour is fashioned in its every form through learning from others. Certain basic expressions, such as smiling, crying and so forth are clearly fundamental to all humans. However, in

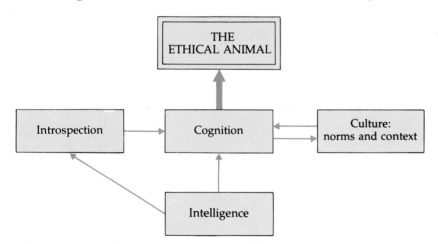

The transcendental complexity of human thought—shaped by our extreme intelligence, capacity for introspection and the cultural context in which we live—gives rise to systems of ethics unique in the living world. Conrad Waddington called *Homo sapiens sapiens* the Ethical Animal.

least specialized of species. Our diet is broad and our geographical distribution world-wide. Through the elaboration of various forms of artifacts, we can fly, dive beneath the oceans and burrow under ground. Humans, in other words, are supremely adaptable, and this has been a prime feature in the success of the species.

Indeed, there are species whose life styles are relatively unspecialized, and there are those in which learning is an essential element of their social and economic activities—a kind of culture. However, no creature approaches the extreme degree of development of these characteristics that is so much a part of

company with many biologists and anthropologists, Geertz believes that the greater part of human behaviour is not tightly constrained by genetics.

An alternative view of human behaviour sees details of cultural expression as a rather flimsy cloak covering underlying genetic imperatives. The chief proponent of this view is Harvard biologist Edward O. Wilson, whose 1975 book *Sociobiology* is a landmark, albeit a rather controversial one, in the study of human behaviour. Since 1975, Wilson has ventured into the debate with several other important publications, chiefly with physicist Charles Lumsden. The two envisage a kind of coevolution of genes and culture, the upshot of which is that virtually every wrinkle of human behaviour is, in part, explicable in terms of genetic determination. They are not suggesting that humans are behavioural automatons marching to the tune of their genes, but that our actions are a great deal more directed than we might like to think.

For example, Wilson and Lumsden consider it more than mildly interesting that categorization of colour perception is universal: most cultures recognize four basic colours—red, yellow, green and blue—which arrangement matches the four types of colour receptors in the retina. They also note the near

This version of the nature–nurture debate has generated a great amount of emotion among protagonists, for many different reasons. It would seem unlikely that human evolution, unusual though it has been, would have stripped its product, *Homo sapiens sapiens*, of all genetically-directed behaviours And yet, the tremendous human adaptability, displayed both by whole populations through the ages and by individuals who move from one social and economic setting to another quite different one, encourages a belief in the supremacy of cultural learning over genetic determinism.

The dramatic expansion of intellectual powers

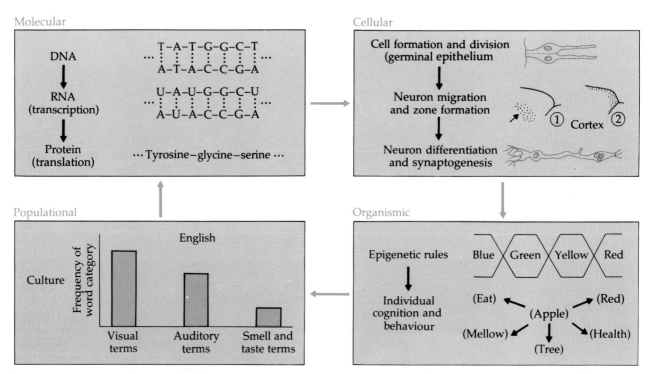

The coevolutionary circuit. Edward O. Wilson and Charles Lumsden envisage a tight relationship between genes and culture (behaviour). In the coevolutionary circuit the genes specify neuronal connections which in turn constrain possible patterns of perception and behaviour. An individual's behavioural choices within a population are constrained by the cultural context, which acts as a filter

(through natural selection) back to individuals in later generations. The coevolutionary circuit—molecular → cellular → organismic → populational → molecular— hones the genetic background to the cultural context. From Wilson E.O. and Lumsden C. (1981) *Genes, Mind and Culture*. Harvard University Press.

universal practice of incest avoidance and the irrational yet powerful emotions engendered by phobias to dangers faced by humans in a more primitive state. Female subjugation by males, acquisitiveness, altruism and many more aspects of human life, all are said to have counterparts in the animal world and all are therefore genetically outlined in the human behavioural repertoire, if not exactly programmed in it.

during human evolution brought with it a degree of introspective consciousness unmatched in the animal world and a capacity, indeed a need, to ethicize. The late Conrad Waddington called man 'the ethical animal', and the capacity to ethicize has been termed the keystone of the specifically human mentality. The ethics and values adopted by any particular group are, of course, inventions of that group. Arbitrary rules, on matters great and trivial, are what provide the structure and meaning to human lives. Echoes of those rules can be seen to emerge in the structure and stylistic conformity imposed on stone tool technologies in the archaeological record; however we can only speculate on what other rules existed in those early societies.

Introspective consciousness is, at the same time, the most personal and yet the most shared of human attributes. By looking into ourselves we can see and understand others, a capacity vital to the functioning of societies as utterly complex as those elaborated by *Homo sapiens sapiens*. It is also the capacity that prompts to ask in humility, Why are we here? And it tempts us into arrogantly placing ourselves above the rest of the animal world, the superior product of evolution.

'We are but a tiny twig on a tree that includes a million species of animals', Stephen Jay Gould reminds us, 'but our one great evolutionary invention, consciousness—a natural product of evolution integrated with a bodily frame of no special merit—has transformed the surface of our planet. Gaze upon the land from an aeroplane window. Has any other species of animal ever left so many visible signs of its relentless presence?'

Those 'visible signs' include, unfortunately, the bad with the good. 'Man is only one of the earth's "manifold creatures"', urges George Gaylord Simpson, 'and he cannot understand his own nature or seek wisely to guide his destiny without taking account of the whole pattern of life.' How will mankind face up to this responsibility? 'I see no reason for despair', responds Simpson, 'But a good deal of reason for pessimism.'

Further reading

Popular discourses on human evolution
Johanson D. & Maitland E. (1981) *Lucy: the Beginnings of Mankind.* Simon & Schuster, New York.
Leakey R. & Lewin R. (1978) *People of the Lake.* Doubleday, New York.
Pfeiffer J. (1978) *The Emergence of Man.* Harper & Row, New York.

A little more detail on the bones and stones
Jolly C. & Plog F. (1978) *Physical Anthropology and Archaeology.* Knopf, New York.
Haddingham E. (1979) *Secrets of the Ice Age.* Walker & Co., London.

Evolutionary biology
Futuyma D. (1979) *Evolutionary Biology.* Sinauer, Sunderland, Mass.
Gould S.J. (1977) *Ever Since Darwin.* W.W. Norton, New York.
Gould S.J. (1980) *The Panda's Thumb.* W.W. Norton, New York.
Gould S.J. (1983) *Hens' Teeth and Horses Toes.* W.W. Norton, New York.
(In addition, Gould's monthly column in *Natural History* magazine is a constant source of dicourse on evolutionary ideas, ancient and modern.)
Jacob F. (1982) *The Possible and the Actual.* Pantheon, New York.
Simpson G.G. (1983) *Fossils and the History of Life.* Scientific American Library, New York.
Stanley S. (1981) *The New Evolutionary Timetable.* Basic Books, New York.

Geology
Press F. & Siever R. (1982) *Earth.* W.M. Freeman, New York.
Redfern R. (1983) *The Making of a Continent.* Times Books, New York.

Genes and culture
Geertz C. (1973) *The Interpretation of Cultures.* Basic Books, New York.
Wilson E.O. (1978) *On Human Nature.* Harvard University Press. Cambridge, Mass.

Molecular biology in palaeoanthropology
Gribbin J. & Cherfas J. (1982) *The Monkey Puzzle.* Pantheon, New York.

Historical perspective
Darwin, Charles. *On the Origin of Species. The Descent of Man, and Selection in Relation to Sex.* Both are available in paperback, usually with helpful introductions by modern scholars.
Huxley, Thomas Henry. *Evidences as to Man's Place in Nature.* Available in several paperback editions.

Index

Note: References to aspects of evolution and *Homo sapiens sapiens* are not entered under these titles but form main headings (or subheadings); a few selected items are entered under Human. **Bold type** indicates the main reference to a topic or name.